Rational Decisions

The Gorman Lectures in Economics

Series Editor, Richard Blundell

A series statement appears at the back of the book

Rational Decisions

Ken Binmore

Princeton University Press
Princeton and Oxford

ISBN: 978-0-691-13074-3 (alk. paper)

Library of Congress Control Number: 2008938513

British Library Cataloging-in-Publication Data is available

This book has been composed in LucidaBright

Typeset by T&T Productions Ltd, London

Printed on acid-free paper ∞

press.princeton.edu

Printed in the United States of America

10 9 8 7 6 5 4 3 2

This book began with two lectures I gave in 2006 at University College London in memory of

Terence Gorman

who was a great economist and a good friend. He would have wished me to dedicate this book to the fondly remembered teacher at his grammar school in Northern Ireland who first inspired him with a love of mathematics.

Contents

Preface

What is rationality? What is the solution to the problem of scientific induction? I don't think it reasonable to expect sharp answers to such questions. One might as well ask for precise definitions of life or consciousness. But we can still try to push forward the frontier of rational decision theory beyond the Bayesian paradigm that represents the current orthodoxy.

Many people see no need for such an effort. They think that Bayesianism already provides the answers to all questions that might be asked. I believe that Bayesians of this stamp fail to understand that their theory applies only in what Jimmie Savage (1951) called a small world in his famous *Foundations of Statistics*. But the world of scientific inquiry is large—so much so that scientists of the future will look back with incredulity at a period in intellectual history when it was possible be taken seriously when claiming that Bayesian updating is the solution to the problem of scientific induction.

Jack Good once claimed to identify 46,656 different kinds of Bayesians. My first priority is therefore to clarify what I think should be regarded as the orthodoxy on Bayesian decision theory—the set of foundational assumptions that offer the fewest hostages to fortune. This takes up most of the book, since I take time out to review various aspects of probability theory along the way. My reason for spending so much time offering an ultra-orthodox review of standard decision theory is that I feel the need to deny numerous misapprehensions (both positive and negative) about what the theory really says—or what I think it ought to say—before getting on to my own attempt to extend a version of Bayesian decision theory to worlds larger than those considered by Savage (chapter 9).

I don't for one moment imagine that my extension of Bayesian decision theory comes anywhere near solving the problem of scientific induction, but I do think my approach will sometimes be found useful in applications. For example, my theory allows the mixed strategies of game theory to be extended to what I call muddled strategies (much as pure strategies were extended to mixed strategies by the creators of the theory).

What is the audience for this book? I hope that it will be read not just by the economics community from which I come myself, but also by statisticians and philosophers. If it only succeeds in bridging some of the gaps between these three communities, it will have been worth-while. However, those seeking a survey of all recent research will need to look for a much bigger book than this. I have tried to include references to literatures that lie outside its scope, but I never stray very far from my own take on the issues. This streamlined approach means that the book may appeal to students who want to learn a little decision theory without being overwhelmed by masses of heavy mathematics or erudite philosophical reasoning, as well as to researchers in the foundations of decision theory.

Sections in which I haven't succeeded in keeping the mathematics at a low level, or where the going gets tough for some other reason, are indicated with an arrow pointing downward in the margin. When such an arrow appears, you may wish to skip to the next full section.

Finally, I want to acknowledge the debt that everyone working on decision theory owes to Duncan Luce and Howard Raiffa, whose *Games and Decisions* remains a source of inspiration more than fifty years after it was written. I also want to acknowledge the personal debt I owe to Francesco Giovannoni, Larry Samuelson, Jack Stecher, Peter Wakker, and Zibo Xu for their many helpful comments on the first draft of this book.

Rational Decisions

1

Revealed Preference

1.1 Rationality?

A rational number is the ratio of two whole numbers. The ancients thought that all numbers were rational, but Pythagoras's theorem shows that the length of the diagonal of a square of unit area is irrational. Tradition holds that the genius who actually made this discovery was drowned, lest he shake the Pythagorean faith in the ineffable nature of number. But nowadays everybody knows that there is nothing irrational about the square root of two, even though we still call it an irrational number.

There is similarly nothing irrational about a philosopher who isn't a rationalist. Rationalism in philosophy consists of arriving at substantive conclusions without appealing to any data. If you follow the scientific method, you are said to be an empiricist rather than a rationalist. But only creationists nowadays feel any urge to persecute scientists for being irrational.

What of rational decision theory? Here the controversy over what should count as rational is alive and kicking.

Bayesianism. Bayesianism is the doctrine that Bayesian decision theory is always rational. The doctrine entails, for example, that David Hume was wrong to argue that scientific induction can't be justified on rational grounds. Dennis Lindley (1988) is one of many scholars who are convinced that Bayesian inference has been shown to be the only coherent form of inference.

The orthodoxy promoted by Lindley and others has become increasingly claustrophobic in economics, but Gilboa and Schmeidler (2001) have shown that it is still possible to consider alternatives without suffering the metaphorical fate of the Pythagorean heretic who discovered the irrationality of $\sqrt{2}$. Encouraged by their success, I follow their example by asking three questions:

What is Bayesian decision theory?

When should we count Bayesian decision theory as rational?

What should we do when Bayesian decision theory isn't rational?

In answering the first question, I hope to distinguish Bayesian decision theory from Bayesianism. We can hold on to the virtues of the former without falling prey to the excesses of the latter.

In answering the second question, I shall note that Leonard (Jimmie) Savage—normally acknowledged as the creator of Bayesian decision theory—held the view that it is only rational to apply Bayesian decision theory in small worlds. But what is a small world?

The worlds of macroeconomics and high finance most certainly don't fall into this category. What should we do when we have to make decisions in such large worlds? I am writing this book because I want to join the club of those who think they have the beginnings of an answer to this third question.[1]

No formal definition of rationality will be offered. I don't believe in the kind of Platonic ideal that rationalist philosophers seem to have in mind when they appeal to Immanuel Kant's notion of Practical Reason. I think that rationality principles are invented rather than discovered. To insist on an a priori definition would be to make the Pythagorean mistake of prematurely closing our minds to possible future inventions. I therefore simply look for a minimal extension of orthodox decision theory to the case of large worlds without pretending that I have access to some metaphysical hotline to the nature of absolute truth.

1.2   Modeling a Decision Problem

When Pandora makes a decision, she chooses an action from those available. The result of her action will usually depend on the state of the world at the time she makes her decision. For example, if she chooses to step into the road, her future fate will depend on whether a car happens to be passing by.

We can capture such a scenario by modeling a decision problem as a function

$$D : A \times B \to C$$

in which A is the set of available actions, B is the set of possible states of the world, and C is the set of possible consequences. So if Pandora chooses action a when the world is in state b, the outcome will be $c = D(a, b)$. Figure 1.1 illustrates a simple case.

An *act* in such a setting is any function $\alpha : B \to C$. For example, if Pandora bets everything she owns on number 13 when playing roulette,

---

[1] For some overviews, see Hammond (1999), Kadane, Schervish, and Seidenfeld (1999), and Kelsey (1992).

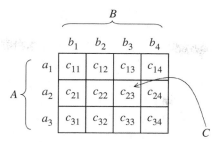

Figure 1.1. A decision problem. Pandora chooses one of the actions: a_1, a_2, a_3. Nature chooses one of the states: b_1, b_2, b_3, b_4. The result is a consequence $c_{ij} = D(a_i, b_j)$. The rows of the matrix of consequences therefore correspond to acts and the columns to states of the world.

then she chooses the act in which she will be wealthy if the little ball stops in the slot labeled 13, and ruined if it doesn't. Pandora's choice of an action always determines some kind of act, and so we can regard A as her set of feasible alternatives within the set $\aleph$ of all possible acts.[2]

If Pandora chooses action a from her feasible set A in a rational way, then we say that a is an optimal choice. The framework we have chosen therefore already excludes one of the most common forms of irrationality—that of choosing an optimal action without first considering what is feasible.

Knowledge. Ken Arrow (1971, p. 45) tells us that each state in B should be "a description of the world so complete that, if true and known, the consequences of every action would be known." But how does Pandora come to be so knowledgeable?

If we follow the philosophical tradition of treating knowledge as justified true belief, the answer is arguably never. I don't see that we are even entitled to assume that reality accords to some model that humans are able to envisage. However, we don't need a view on the metaphysical intelligibility of the universe to discuss decision theory intelligently. The models we use in trying to make sense of the world are merely human inventions. To say that Pandora knows what decision model she is facing can therefore be taken as meaning no more than that she is committed to proceeding as though her model were true (section 8.5).

1.3 Reason Is the Slave of the Passions

Thomas Hobbes characterized man in terms of his strength of body, his passions, his experience, and his reason. When Pandora is faced with a

---

[2] The Hebrew letter $\aleph$ is pronounced "aleph."

decision problem, we may identify her strength of body with the set A of all actions that she is physically able to choose. Her passions can be identified with her preferences over the set C of possible consequences, and her experience with her beliefs about the likelihood of the different possible states in the set B. In orthodox decision theory, her reason is identified with the manner in which she takes account of her preferences and beliefs in deciding what action to take.

The orthodox position therefore confines rationality to the determination of means rather than ends. To quote David Hume (1978): "Reason is, and ought only to be, the slave of the passions." As Hume extravagantly explained, he would be immune to accusations of irrationality even if he were to prefer the destruction of the entire universe to scratching his finger. Some philosophers hold to the contrary that rationality can tell you what you ought to like. Others maintain that rationality can tell you what you ought to choose without reference to your preferences. For example, Kant (1998) tells us that rationality demands that we honor his categorical imperative, whether or not we like the consequences.

My own view is that nothing is to be gained in the long run by inventing versions of rationality that allow their proponents to label brands of ethics or metaphysics other than their own as irrational. Instead of disputing whose ethical or metaphysical system should triumph, we are then reduced to disputing whose rationality principles should prevail. I prefer to emulate the logicians in seeking to take rationality out of the firing line by only adopting uncontroversial rationality principles.

Such a minimalist conception of rational decision theory isn't very glamorous, but then, neither is modern logic. However, as in logic, there is a reward for following the straight and narrow path of rectitude. By so doing, we will be able to avoid getting entangled in numerous thorny paradoxes that lie in wait on every side.

Consistency.   The modern orthodoxy goes further than David Hume. It treats reason as the slave, not only of our passions, but also of our experience. Pandora's reason is assumed to be the slave of both her preferences and her beliefs.

It doesn't follow that rational decision theory imposes no constraints on our preferences or our beliefs. Everyone agrees that rational people won't fall prey to the equivalent of a logical contradiction. Their preferences and beliefs will therefore be consistent with each other in different situations. But what consistency criteria should we impose? We mustn't be casual about this question, because the words *rationality* and *consistency* are treated almost as synonyms in much modern work.

rara ve3

For example, Bayesianism focuses on how Pandora should respond to a new piece of data. It is said that Pandora must necessarily translate her prior beliefs into posterior beliefs using Bayes' rule if she is to act consistently. But there is seldom any serious discussion of *why* Pandora should be consistent in the sense required. However, this is a topic for a later chapter (section 7.5.2). We already have enough contentious issues in the current chapter to keep us busy for some time.

1.4 Lessons from Aesop

deteriorarse ?
fábula
así

The fox in Aesop's fable was unable to reach some grapes and so decided that they must be sour. He thereby irrationally allowed his beliefs in domain *B* to be influenced by what actions are feasible in domain *A*.

If Aesop's fox were to decide that chickens must be available because they taste better than grapes, he would be guilty of the utopian mistake of allowing his assessment of what actions are available in domain *A* to be influenced by his preferences in domain *C*. The same kind of wishful thinking may lead him to judge that the grapes he can reach must be ripe because he likes ripe grapes better than sour grapes, or that he likes sour grapes better than ripe grapes because the only grapes that he can reach are probably sour. In both these cases, he fails to separate his beliefs in domain *B* from his preferences in domain *C*.

Aesop's principle.   These observations motivate the following principle:

> Pandora's preferences, her beliefs, and her assessments of what is feasible should all be independent of each other.

For example, the kind of pessimism that might make Pandora predict that it is bound to rain now that she has lost her umbrella is irrational. Equally irrational is the kind of optimism that Voltaire was mocking when he said that if God didn't exist, it would be necessary to invent Him.

1.4.1 Intrinsic Preferences?

It is easy to propose objections to Aesop's principle. For example, Pandora's preferences between an umbrella and an ice cream might well alter if it comes on to rain. Shouldn't we therefore accept that preferences will sometimes be state-dependent?

It is true that *instrumental* preferences are usually state-dependent. One can tell when one is dealing with an instrumental preference, because it advances matters to ask Pandora *why* she holds the preference. She might say, for example, that she prefers an umbrella to an ice cream because it looks like rain and she doesn't want to get wet. Or that she prefers driving nails home with a hammer rather than a screwdriver because it takes less time and trouble. More generally, any preference over actions is an instrumental preference.

One is dealing with *intrinsic* preferences when it no longer helps to ask Pandora why she likes one thing rather than another, because nothing relevant that might happen is capable of altering her position.[3] For example, we could ask Pandora why she likes wearing a skirt that is two inches shorter than the skirts she liked last year. She might reply that she likes being in the fashion. Why does she like being in the fashion? Because she doesn't like being laughed at for being behind the times. Why doesn't she like being laughed at? Because girls who are ridiculed are less attractive to boys. Why does she like being attractive to boys? One could reply that evolution made most women this way, but such an answer doesn't take us anywhere, because we don't plan to consider alternative environments in which evolution did something else. The fact that Pandora likes boys can therefore usefully be treated as an intrinsic preference. Her liking for miniskirts is instrumental because it changes with her environment.

Economists are talking about intrinsic preferences when they quote the slogan: *De gustibus, non est disputandum.* In welfare economics, it is particularly important that the preferences that we seek to satisfy should be intrinsic. It wouldn't help very much, for example, to introduce a reform that everyone favors if it changes the environment in a way that reverses everyone's preferences.

1.4.2 Constructing a Decision Problem

Decision problems aren't somehow built into the structure of the universe. Pandora must decide how to formulate her decision problem for herself. There will often be many formulations available, some of which satisfy the basic assumptions of whatever decision theory she plans to apply—and others which don't. She might have to work hard, for example, to find a formulation in which her preferences on the set C of consequences are intrinsic. If she plans to apply Aesop's principle completely, she must work even harder and construct a model in which there are no linkages at all between any of the sets A, B, and C (other than those built into the function D) that are relevant to her decision.

For example, if Pandora's actions are predictions of the weather, she mustn't take the states in B to be *correct* or *mistaken*, because the true

---

[3] The distinction between intrinsic preferences and instrumental preferences is made in economics by speaking of direct and indirect utility functions.

state of the world would then depend on her choice of action. In the case of an umbrella on a rainy day, it may be necessary to identify the set C of consequences with Pandora's *states of mind* rather than physical objects. One can then speak of the states of mind that accompany having an umbrella-on-a-sunny-day or having an umbrella-on-a-wet-day, rather than speaking just of an umbrella.

Critics complain that the use of such expedients makes the theory tautological, but what could be better when one's aim is to find an uncontroversial framework?

Every thing is what it is, and not something else. The price that has to be paid for an uncontroversial theory is that it can't be used to model everything that we might like to model. For example, we can't model the possibility that people might choose to change their intrinsic preferences by adopting behaviors that are likely to become habituated. Such restrictions are routinely ignored when appeals are made to rational decision theory, but to bowdlerize Bishop Butler: Every theory is good for what it is good for, and not for something else.

To ignore the wisdom of Bishop Butler is to indulge in precisely the kind of wishful thinking disbarred by Aesop's principle. We aren't even entitled to take for granted that Pandora *can* formulate her decision problem so that Aesop's principle applies. In fact, I shall be arguing later that it is a characteristic of decision making in large worlds that agents are *unable* to separate their preferences from their beliefs. However, this unwelcome consideration will be left on ice until chapter 9.

Finally, nothing says that one can't construct a rational decision theory that applies to decision problems that don't satisfy Aesop's principle. Richard Jeffrey (1965) offers a theory in which your own choice of action may provide evidence about the choices made by other people. Ed Karni (1985) offers a theory in which the consequences may be inescapably state-dependent. I am hesitant about such theories because I don't know how to evaluate their basic assumptions.

1.5 Revealed Preference

The theory of revealed preference goes back a long way. Thomas Hobbes (1986), for example, sometimes writes as though choosing a when b is available is unproblematically the same as liking a better than b. However, its modern incarnation is usually attributed to the economist Paul Samuelson (1947), although the basic idea goes back at least as far as Frank Ramsey (1931).

So much confusion surrounds the simple idea of revealed preference that it is worth reviewing its history briefly. The story begins with Jeremy Bentham's adoption of the term *utility* as a measure of the pleasure or pain a person feels as a result of a decision being made. Perhaps he thought that some kind of metering device might eventually be wired into Pandora's brain that would show how many units of utility (utils) she was experiencing. This is a less bold hypothesis than it may once have seemed, since we now know that rats will pull a lever that activates an electrode embedded in a pleasure center in their brains in preference to anything else whatever—including sex and food. It is therefore unsurprising that a modern school of behavioral economists have reverted to this classical understanding of the nature of utility. A more specialized group devote their attention specifically to what they call happiness studies. However, the theory of revealed preference remains the orthodoxy in economic theory (Mas-Collel et al. 1995).

1.5.1   Freeing Economics from Psychology

Economists after Bentham became increasingly uncomfortable, not only with the naive hypothesis that our brains are machines for generating utility, but with all attempts to base economics on psychological foundations. The theory of revealed preference therefore makes a virtue of assuming *nothing whatever* about the psychological causes of our choice behavior.

This doesn't mean that economists believe that our choice behavior isn't caused by what goes on in our heads. Adopting the theory of revealed preference doesn't entail abandoning the principle that reason is the slave of the passions. Studies of brain-damaged people show that when our capacity for emotional response is impaired, we also lose our capacity to make coherent decisions (Damasio 1994). Nor is there any suggestion that we all make decisions in the same way. The theory of revealed preference accepts that some people are clever, and others are stupid; that some care only about money, and others just want to stay out of jail. Nor does the theory insist that people are selfish, as its critics mischievously maintain. It has no difficulty in modeling the kind of saintly folk who would sell the shirt off their back rather than see a baby cry.

Modern decision theory succeeds in accommodating the infinite variety of the human race within a single theory simply by denying itself the luxury of speculating about what is going on inside someone's head. Instead, it pays attention only to what people do. It assumes that we

already know what people choose in some situations, and uses this data
to deduce what they will choose in other situations.

For example, Pandora may buy a bundle of goods on each of her weekly
visits to the supermarket. Since her household budget and the supermarket prices vary from week to week, the bundle she purchases isn't always
the same. However, after observing her shopping behavior for some time,
one can make an educated guess about what she will buy next week, once
one knows what the prices will be, and how much she will have to spend.

Stability. In making such inferences, two assumptions are implicitly
understood. The first is that Pandora's choice behavior is *stable*. We won't
be able to predict what she will buy next week if something happens
today that makes our data irrelevant. If Pandora loses her heart to a
football star, who knows how this might affect her shopping behavior?
Perhaps she will buy no pizza at all, and start filling her shopping basket
with deodorant instead.

Amartya Sen (1993) offers a rather different example based on the
observation that people never take the last apple from a bowl, but will
take an apple from the bowl when it contains two apples. Are their preferences over apples therefore unstable, in that sometimes they like apples
and sometimes they don't?

Sen's example underlines the care that Pandora must exercise in formulating her decision problem. The people in Sen's story inhabit some
last bastion of civilization where Miss Manners still reigns supreme—
and this is relevant when Pandora comes to model her problem. Her
belief space B must allow her to recognize that she is taking an apple
from a bowl in a society that subscribes to the social values of Miss
Manners rather than those of Homer Simpson. Her consequence space
C must register that the subtle social punishments that her fellows will
inflict on her for deviating from their social mores in taking the last
apple from a bowl are at least as important to her as whether she gets
to eat an apple right now. Pandora's apparent violation of the stability requirement of revealed-preference theory then ceases to bite. Her
choice behavior reveals that she likes apples enough to take one when
no breach of etiquette is involved, but not otherwise.

1.5.2 Consistency

The second implicit assumption required by the theory of revealed preference is that Pandora's choice behavior must be *consistent*. We certainly
won't be able to predict what she will do next if she just picks items off
supermarket shelves at random, whether or not they are good value or

satisfy her needs. We are therefore back with the question: What criteria
determine whether Pandora's choice behavior is consistent?

Strict preferences.   It is usual to write $a \prec b$ to mean that Pandora likes
b strictly more than a. Such a strict preference relation is said to be
consistent if it is both asymmetric and transitive.

   A preference relation $\prec$ is transitive if

$$a \prec b \text{ and } b \prec c \quad \text{implies} \quad a \prec c. \tag{1.1}$$

It is only when transitivity holds that we can describe Pandora's preferences by simply writing $a \prec b \prec c$. Without transitivity, this information
wouldn't imply that $a \prec c$.

   A preference relation $\prec$ is asymmetric if we don't allow both $a \prec b$
and $b \prec a$. It represents a full set of strict preferences on a set X if we
insist that either $a \prec b$ or $b \prec a$ must always hold for any a and b in X
that aren't equal.[4]

   Given a set of full and consistent strict preferences for Pandora over
a finite set X, we can predict the unique alternative $x = y(A)$ that she
will choose from each subset A of X. We simply identify $y(A)$ with the
alternative x in A for which $a \prec x$ for all other alternatives a in A.

Reversing the logic.   Instead of constructing a choice function y from
a given preference relation $\prec$, a theory of revealed preference needs to
construct a preference relation from a given choice function.

   If we hope to end up with a *strict* preference relation, the choice function y will need to assign a *unique* optimal outcome $y(A)$ to each subset
A of a given finite set X. We can then define $\prec$ by saying that Pandora
prefers b to a if and only if she chooses b from the set $\{a, b\}$. That is to
say:

$$a \prec b \quad \text{if and only if} \quad b = y\{a, b\}. \tag{1.2}$$

A strict reference relation constructed in this way is automatically asymmetric, but it need not be transitive. To obtain a consistent preference
relation, we need the choice function to be consistent in some sense.

Independence of Irrelevant Alternatives.   There is a large economic literature (Richter 1988) on consistent choice under market conditions which
I propose to ignore. In the absence of information about prices and bud-
gets, the theory of revealed preference has much less bite, but is correspondingly simpler. The only consistency requirement we need right

---

[4] The technical term is *complete* or *total*.

now is the Independence of Irrelevant Alternatives:[5]

> If Pandora sometimes chooses b when a is feasible, then she never chooses a when b is feasible.

The following argument by contradiction shows that this consistency requirement implies that the revealed-preference relation defined by (1.2) is transitive.

If the transitivity requirement (1.1) is false for some a, b, and c, then $b = y\{a, b\}$, $c = y\{b, c\}$, and $a = y\{c, a\}$. The fact that $b = y\{a, b\}$ shows that b is sometimes chosen when a is feasible. It follows that a can't be chosen from $\{a, b, c\}$ because b is feasible. Similar arguments show that b and c can't be chosen from $\{a, b, c\}$ either. But Pandora must choose one of the alternatives in $\{a, b, c\}$, and so we have a contradiction.

Another example from Amartya Sen (1993) will help to explain why Pandora may have to work hard to find a formulation of her decision problem in which a consistency requirement like the Independence of Irrelevant Alternatives holds.

A respectable lady is inclined to accept an invitation to take tea until she is told that she will also have the opportunity to snort cocaine. She thereby falls foul of the Independence of Irrelevant Alternatives. There are no circumstances in which she would choose to snort cocaine, and so removing the option from her feasible set should make no difference to what she regards as optimal.

The reason that she violates the Independence of Irrelevant Alternatives without our finding her behavior unreasonable is that snorting cocaine isn't an *irrelevant* alternative for her. The fact that cocaine is on the menu changes her beliefs about the kind of person she is likely to meet if she accepts the invitation. If we want to apply the theory of revealed preference to her behavior, we must therefore find a way to formulate her decision problem in which no such hidden relationships link the actions available in her feasible set A with either her beliefs concerning the states in the set B or the consequences in the set C.

Indifference.   The preceding discussion was simplified by only looking at cases in which all of Pandora's preferences are strict. Taking account of the possibility of indifference is a bit of a nuisance, but we can deal with the problem very quickly.

If Pandora is sometimes indifferent, it is necessary to abandon the assumption that $\prec$ fully describes her preferences over X. However, we

---

[5] The Independence of Irrelevant Alternatives is used here in the original sense of Nash (1950). It is unfortunate that Arrow (1963) borrowed the same terminology for a related but different idea.

can define a full relation $\preceq$ on X by

$$a \preceq b \quad \text{if and only if} \quad \text{not}(b \prec a).$$

We say that $\preceq$ is a weak preference relation if it is transitive.[6] We don't need to postulate that the strict preference relation $\prec$ is transitive separately, because this follows from our other assumptions. If $\preceq$ is a weak preference relation, we write $a \sim b$ if and only if $a \preceq b$ and $b \preceq a$. Pandora is then said to be indifferent between a and b.

A weak preference relation $\preceq$ on X doesn't necessarily determine a unique choice for Pandora in each subset A of X because she may be indifferent among several alternatives, all of which are optimal. So the notation $y(A)$ now has to denote the choice *set* consisting of all x in A for which $a \preceq x$ for all a in A. Pandora is assumed to regard each alternative in $y(A)$ as an equally good solution of her decision problem.

When we reverse the logic by seeking to deduce a weak preference relation $\preceq$ from a choice function y, we must therefore allow the value of $y(A)$ to be any nonempty subset of A. We can still define $\prec$ as in (1.2), but now

$$a \sim b \quad \text{if and only if} \quad \{a, b\} = y\{a, b\}.$$

If the relation $\preceq$ constructed in this way is to be a weak preference relation, we need to impose some consistency requirement on Pandora's choices to ensure that $\preceq$ is transitive. It suffices to strengthen the Independence of Irrelevant Alternatives to Houthakker's axiom:[7]

> If Pandora sometimes includes b in her choice set when a is feasible, then she never includes a in her choice set when b is feasible without including b as well.

1.6 Rationality and Evolution

Evolution is about the survival of the fittest. Entities that promote their fitness consistently will therefore survive at the expense of those that promote their fitness only intermittently. When biological evolution has had a sufficiently long time to operate, it is therefore likely that each relevant locus on a chromosome will be occupied by the gene with maximal fitness. Since a gene is just a molecule, it can't *choose* to maximize its fitness, but evolution makes it seem as though it had. This is a valuable insight, because it allows biologists to use rationality considerations

---

[6] Which means that (1.1) holds with $\prec$ replaced by $\preceq$.

[7] I follow David Kreps (1988) in attributing the axiom to Houtthaker, but it seems to originate with Arrow (1959).

to predict the outcome of an evolutionary process, without needing to follow each complicated twist and turn that the process might take.

When appealing to rationality in such an evolutionary context, we say that we are seeking an explanation in terms of *ultimate* causes rather than *proximate* causes. Why, for example, do songbirds sing in the early spring? The proximate cause is long and difficult. This molecule knocked against that molecule. This chemical reaction is catalyzed by that enzyme. But the ultimate cause is that the birds are signalling territorial claims to each other in order to avoid unnecessary conflict. They neither know nor care that this behavior is rational. They just do what they do. But the net effect of an immensely complicated evolutionary process is that songbirds behave *as though* they had rationally chosen to maximize their fitness.

Laboratory experiments on pigeons show that they sometimes honor various consistency requirements of rational choice theory better than humans (Kagel, Battalio, and Green 1995). We don't know the proximate explanation. Who knows what goes on inside the mind of a pigeon? Who knows what goes on in the minds of stockbrokers for that matter? But we don't need to assume that pigeons or stockbrokers are thinking rationally because we see them behaving rationally. We can appeal to the ultimate explanations offered by evolutionary theory.

People naturally think of biology when such appeals to evolution are made, but I follow the practice in economics of using the term more widely. After Alchian (see Lott 1997), it is common to argue that the forces of social or economic evolution will tend to eliminate stockbrokers who don't consistently seek to maximize profit. It is therefore unsurprising that evolutionary arguments are sometimes marshaled in defense of rationality concepts. The money pump argument is a good example.

Money pumps.   Why should we expect Pandora to reveal transitive preferences? The money pump argument says that if she doesn't, other people will be able to make money out of her.

Suppose that Pandora reveals the intransitive preferences

$$\text{apple} \prec \text{orange} \prec \text{fig} \prec \text{apple}$$

when making pairwise comparisons. A trader now gives her an apple. He then offers to exchange the apple for an orange, provided she pays him a penny. Since her preference for the orange is strict, she agrees. The trader now offers to exchange the orange for a fig, provided she pays him a penny. When she agrees, the trader offers to exchange the fig for an apple, provided she pays him a penny.

If Pandora's preferences are stable, this trading cycle can be repeated until Pandora is broke. The inference is that nobody with intransitive preferences will be able to survive in a market context.

1.7 Utility

In the theory of revealed preference, utility functions are no more than a mathematical device introduced to help solve choice problems. A preference relation $\preceq$ is represented by a real-valued utility function u if and only if

$$u(a) \leqslant u(b) \quad \text{if and only if} \quad a \preceq b.$$

Finding an optimal x then reduces to solving the maximization problem:

$$u(x) = \max_{a \in A} u(a), \tag{1.3}$$

for which many mathematical techniques are available.

Constructing utility functions. Suppose that Pandora's choice behavior reveals that she has consistent preferences over the five alternatives a, b, c, d, and e. Her revealed preferences are

$$a \prec b \sim c \prec d \prec e.$$

It is easy to find a utility function U that represents Pandora's preferences. She regards the alternatives a and e as the worst and the best available. We therefore set $U(a) = 0$ and $U(e) = 1$. We next pick any alternative intermediate between the worst and the best alternative, and take its utility to be $\frac{1}{2}$. In Pandora's case, b is an alternative intermediate between a and e, and so we set $U(b) = \frac{1}{2}$. Since $b \sim c$, we must also set $U(c) = \frac{1}{2}$. Only the alternative d remains. This is intermediate between c and e, and so we set $U(d) = \frac{3}{4}$, because $\frac{3}{4}$ is intermediate between $U(c) = \frac{1}{2}$ and $U(e) = 1$.

The utilities we have assigned to alternatives are ranked in the same way as the alternatives themselves. In making choices, Pandora therefore behaves *as though* she were maximizing the value of U. But she also behaves as though she were maximizing the values of the alternative utility functions V and W. There are also many other ways that we could have assigned utilities to the alternatives in a manner consistent with Pandora's preferences. In the theory of revealed preference, the only criterion that is relevant when picking one of the infinity of utility functions that represent a given preference relation is that of mathematical convenience.

a	a	b	c	d	e
$U(a)$	0	$\frac{1}{2}$	$\frac{1}{2}$	$\frac{3}{4}$	1
$V(a)$	-1	1	1	2	3
$W(a)$	-8	0	0	1	8

Figure 1.2. Constructing utility functions. The method always works for a consistent preference relation defined over a finite set of alternatives, because there is always another real number between any pair of real numbers.

1.7.1 Cardinal versus Ordinal

The distinction between ordinal and cardinal notions of utility is left over from a controversy of the early twentieth century. Jeremy Bentham's identification of utility with pleasure or pain was the orthodox position among early Victorian economists, but a generation of reforming economists led by Stanley Jevons (1871) showed that one can often dispense with appeals to utility altogether by carefully considering what happens "at the margin." In a manner familiar to historians of thought, the generation following the marginalist revolution then poured scorn on what they saw as the errors of their Victorian forebears. The jeremiads of the economist Lionel Robbins (1938) are still quoted by modern critics who are unaware that modern utility theory has totally abandoned the shaky psychological foundations proposed by Bentham and his followers.

Robbins saw no harm in the use of ordinal utility functions as a mathematical convenience. These are utility functions constructed like those of figure 1.2 so that only the *ranking* of their values has any significance. For example, the fact that $U(e) - U(d) = U(d) - U(c)$ doesn't tell us that Pandora would be as willing to swap d for e as she would be to swap c for d. After all, we were free to assign any value to $U(d)$ between $\frac{1}{2}$ and 1. The fact that only the ranking of alternatives is preserved in an ordinal utility function is expressed mathematically by saying that any ordinal utility function is a strictly increasing transformation of any other ordinal utility function that represents the same preference relation. For example, $V(a) = -1 + 4U(a)$ and $W(a) = \{V(a) - 1\}^3$.

One might think of the utility scale created by an ordinal utility function as being marked on an elastic measuring tape. No matter how one squeezes or stretches the tape, the result will still be an ordinal utility

scale. A cardinal utility function is one that creates a scale like temperature. One is free to choose the zero and the unit on such a scale, but then
there is no more room for maneuver. In mathematical terms, any cardinal utility function is only a strictly increasing *affine* transformation of
any other cardinal utility function that represents the same preference
relation.[8]

Robbins drew the line at cardinal utility functions. It is ironic that he
was still denouncing them as intrinsically meaningless long after the
moment at which John Von Neumann explained to Oskar Morgenstern
how to make sense of them in risky situations. However, this is a story
that must wait until section 3.4.

The immediate point is that Robbins was perfectly right to insist that
the mere fact that we might somehow be gifted with cardinal utility
functions doesn't imply that we can necessarily compare Pandora's utils
meaningfully with those of other people. To do so would be like thinking
that a room in which the temperature is 32 °Fahrenheit must be warmer
than a room in which the temperature is 31 °Celsius. If we want to use
utility functions to make interpersonal comparisons of welfare, we certainly need a theory that generates cardinal utility functions rather than
the ordinal utility functions considered in this chapter, but we also need
a lot more input concerning the social context (section 4.4).

1.7.2 Infinite Sets

It is sometimes argued that infinite sets don't actually exist, and so we
need not make our lives complicated by worrying about them. But this
argument completely misses the point. Infinite sets aren't introduced
into models for metaphysical reasons. They are worth worrying about
because infinite models are often much *simpler* than the finite models
for which they serve as idealizations.

Finding a utility representation of a consistent preference relation
defined on an infinite set X is no problem when X is countable.[9] The
method of section 1.6 works equally well in this case, provided we allow
the utility function to take values outside the range determined by the
first pair of alternatives that we choose to consider. If we like, the values
of the utility function so constructed can be made to lie between any two
predetermined bounds.

---

[8] The equation $y = Ax + B$ defines an affine transformation from the real numbers to
the real numbers. It is strictly increasing if and only if $A > 0$. Thus V can be obtained by
applying the strictly increasing affine transformation $y = 4x - 1$ to U.

[9] This means that the members of X can be arranged in a sequence and so be counted.
The set of all n-tuples of rational numbers is a countable dense subset of the uncountable
set $\mathbb{R}^n$ of all n-tuples of real numbers.

Given a consistent preference relation on a set X with a countable dense subset Y, we can first construct a utility representation on Y as above. The representation can then be extended to the whole of X by continuity, provided that all the sets $\{a \in X : a \preceq b\}$ and $\{a \in X : a \succeq b\}$ are closed for all b in X. The resulting utility function will then be continuous, which guarantees that the maximization problem (1.3) has a solution when the feasible set A is compact.[10]

Lexicographic preferences. It isn't true that all consistent preference relations on infinite sets have a utility representation. For example, in the torture example of the next section, Pandora might care primarily about the intensity of pain she must endure, taking account of the time it must be endured only when comparing two situations in which the intensity of pain is the same. It is impossible to represent such a lexicographic preference with a utility function.[11]

Any such representation would assign an open interval of real numbers to each vertical line in figure 1.3. But there are an uncountable number of such verticals, and any collection of nonoverlapping open intervals must be countable (because we can label each such open interval with one of the rational numbers it contains).

1.8 Challenging Transitivity

This section reviews two of the many arguments that continue to be directed against transitivity by philosophers.

Paradox of preference. This example is taken from a recent compendium of paradoxes put together by Clark (2002). You would choose to fly a glider for the first time accompanied by an experienced instructor rather than try out a racing car. But you would choose the racing car rather than flying the glider solo. However, your machismo requires that you choose flying the glider solo to flying with an instructor if that is the only choice on offer.

As with Sen's examples, the problem hasn't been adequately formulated. At the very least, it is necessary to distinguish between doing something without loss of machismo and doing it with loss of machismo.

---

[10] To say that Y is dense in X means that we can approximate any point in X as closely as we like by points in Y. To say that a set is closed means that it contains all its boundary points. To say that a set in $\mathbb{R}^n$ is compact means that it is closed and bounded.

[11] Lexicographic means alphabetical. If Pandora had lexicographic preferences over words, then she would prefer whichever of two words came first in the dictionary.

Achilles and the tortoise.   In arguing that the relationship "all things considered better than" need not be transitive, Stuart Rachels (1998, 2001) and Larry Temkin (1996) make three claims about human experience:[12]

Claim 1. Any unpleasant experience, no matter what its intensity and duration, is better than a slightly less intense experience that lasts much longer.

Claim 2. There is a finely distinguishable range of unpleasant experiences ranging in intensity from mild discomfort to extreme agony.

Claim 3. No matter how long it must be endured, mild discomfort is preferable to extreme agony for a significant amount of time.

These three claims are invoked to justify a story, which begins with our being invited to consider two extreme alternatives. In the first, Pandora suffers an excruciating torture for a short period of time. In the second, she endures a mild pain for a very long time. Between these two extremes lie a chain of alternatives in which the pain keeps being reduced very slightly but the time it must be endured is prolonged so that Pandora never regards the next alternative as better than its predecessor. But Pandora prefers the final alternative to the initial alternative, and so her preferences go round in a circle, which transitivity doesn't allow.

The argument can be shown to be wrong simply by considering the special case in which Pandora's preferences are determined by the utility function

$$u(p, t) = -pt/(1 + t),$$

in which p is the intensity of pain and t the length of time it must be endured. The preferences that Pandora reveals are necessarily transitive, because this is always true of preferences obtained from a utility function. However, Pandora nevertheless satisfies all three of the claims said to imply that her preferences must sometimes be intransitive.

Figure 1.3 shows how utility functions are often illustrated in diagrams by drawing indifference curves along which Pandora's utility is constant. The little arrows show the direction of Pandora's preference at any pair (p, t) representing an intensity p of pain endured for a period of length t. The chain of alternatives in the argument are shown as the sequence $(p_1, t_1), (p_2, t_2), (p_3, t_3), \ldots$. The figure makes it clear that we aren't entitled to assume that the intensity of pain in this sequence can be brought as close to zero as we choose. To assume otherwise is to fall prey to the paradox of Achilles and the tortoise.

---

[12] See Binmore and Voorhoeve (2003, 2006) and Voorhoeve (2007) for a less terse account. Quinn's (1990) paradox, which is similar but substitutes the Sorites paradox for Zeno's paradox, is also considered.

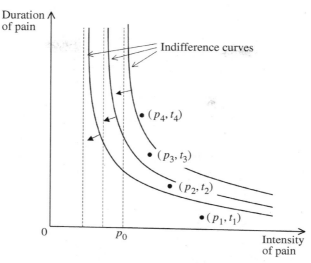

Figure 1.3. Achilles and the tortoise. Pandora likes it less and less as she progresses through the sequence (p_n, t_n), but she never gets to a stage where the pain is negligible, because $p_n > p_0$ for all values of n.

My own view is that such arguments fail to take proper account of the proviso that "all things are to be considered." Pandora should also consider what she would choose if offered feasible sets containing *three* alternatives. If she doesn't violate Houthakker's axiom, the preferences she reveals will then necessarily be transitive.

1.9 Causal Utility Fallacy

The fact that utility used to mean one thing and now means another is understandably the cause of widespread confusion. Bentham himself suggested substituting the word *felicity* for his notion of utility as pleasure or pain, but I suppose it is hopeless to propose that his suggestion be taken up at this late date. The chief misunderstanding that arises from confusing a utility derived from revealed-preference theory with a Benthamite felicity is that the former offers no explanation of why Pandora should choose one thing rather than another, while the latter does.

In revealed-preference theory, it isn't true that Pandora chooses *b* rather than *a* *because* the utility of *b* exceeds the utility of *a*. This is the Causal Utility Fallacy. It isn't even true that Pandora chooses *b* rather than *a* because she prefers *b* to *a*. On the contrary, it is because Pandora chooses *b* rather than *a* that we say that Pandora prefers *b* to *a*, and assign *b* a larger utility.

The price of abandoning psychology for revealed-preference theory is therefore high. We have to give up any pretension to be offering a causal explanation of Pandora's choice behavior in favor of an account that is merely a description of the choice behavior of someone who chooses consistently. Our reward is that we end up with a theory that is hard to criticize because it has little substantive content.

Notice that Pandora might be choosing consistently because she is seeking to maximize the money in her pocket, or consciously pursuing some other end. In an evolutionary context, Pandora might be an animal who isn't choosing consciously at all, but who makes consistent choices because animals who failed to promote their fitness consistently have long since been eliminated. Neither of these possibilities is denied by the theory of revealed preference, which is entirely neutral about *why* Pandora makes rational decisions.

In such cases, we can simply identify utility with money or fitness without any hassle. It is only when no obvious maximand is in sight that the theory of revealed preference comes into its own. It tells us that if Pandora's choice behavior is consistent, then we can model her as maximizing some abstract quantity called utility. She may not be aware that she is behaving as though maximizing the utility function we have invented for her, but she behaves as though she is a maximizer of utility nevertheless.

Neoclassical economics.   The theory of revealed preference is a child of the marginalist revolution. As such, it is an official doctrine of neoclassical economics, enshrined in all respectable textbooks.

However, the kind of critic who thinks that economists are mean-minded, money-grubbing misfits commonly ignores the official doctrine in favor of a straw man that is easier to knock down. It is said that neoclassical economics is based on the principle that people are selfish. Henrich et al. (2004) even invent a "selfishness axiom" that supposedly lies at the heart of economics. Sometimes it is said that economists necessarily believe that people care only about money, so that utility theory reduces to a crude identification of utils with dollars.

It is true that economists are more cynical about human behavior than the general run of humanity. I became markedly more cynical myself after becoming a business consultant to raise money for my research center. It is an unwelcome fact that many people do behave like mean-minded, money-grubbing misfits in the business world. Nor is it true that the laboratory experiments of behavioral economists show that ordinary people are always better behaved. After ten trials or so, nearly all subjects end up defecting in the Prisoners' Dilemma (Ledyard 1995). However, the empirical facts on the ground are irrelevant to the immediate issue.

It isn't true that it is axiomatic in economic theory that people are selfish. I suspect that this widespread misunderstanding arises because people think that rational agents must act out of self-interest because they maximize their own utility functions rather than some social objective function. But to make this claim is to fail to understand the theory of revealed preference. Whatever the explanation of Pandora's behavior, if she acts consistently, she will act as though maximizing a utility function tailored to her own behavior. Tradition is doubtless right that St Francis of Assisi consistently promoted the interests of the poor and sick. He therefore behaved as though maximizing a utility function that took account of their welfare. But nobody would want to say that St Francis was selfish because the particular form of his utility function was idiosyncratic to him.

In the past, I have followed certain philosophers in saying that people act in their own *enlightened* self-interest when they act consistently, but I now see that this was a mistake. Any kind of language that admits the Causal Utility Fallacy as a possibility is best avoided.

Preference satisfaction. Most accounts of rational decision theory in the philosophical literature fall headlong into the Causal Utility Fallacy.[13] For example, Gauthier (1993, p. 186) tells us that "only utilities provide reasons for acting." Accounts that avoid this mistake usually follow A. J. Ayer in talking about "preference satisfaction." Utility functions are indeed constructed from preferences, but preferences themselves aren't primitives in the theory of revealed preference. The primitives of the theory are the *choices* that Pandora is seen (or hypothesized) to make.

Rational choice theory. There is an ongoing debate in political science about the extent to which what they call rational choice theory can sensibly be applied. Neither side in this often heated debate will find much comfort in the views expressed in this chapter. The issue isn't whether people are capable of the superhuman feats of calculation that would be necessary to consciously maximize a utility function in most political contexts. After all, spiders and fish can't be said to be conscious at all, but evolutionary biologists nevertheless sometimes succeed in modeling them as maximizers of biological fitness. As in philosophy, both the

---

[13] Simon Blackburn's (1998, chapter 6) *Ruling Passions* is an interesting philosophical text that gives an accurate account of revealed-preference theory. Like Gibbard (1990), Blackburn seeks proximate causes of our moral behavior rooted in our emotional responses. My own complementary approach looks for ultimate causes in our evolutionary history (Binmore 2005).

proponents and the opponents of rational choice theory need to learn that their theory is based on consistency of *choice* (for whatever reason) rather than on conscious preference satisfaction.

1.10   Positive and Normative

What is rationality good for? The standard answer is that it is good for working out the ideal means of achieving whatever your ends may be. Economists traditionally offer another answer. They attribute rationality to the agents in their models when seeking to predict how ordinary people will actually behave in real situations. Both answers deserve some elaboration.

Normative applications.   A theory is normative or prescriptive if it says what ought to happen. It is positive or descriptive if it predicts what will actually happen.

Jeremy Bentham's approach to utility was unashamedly normative. He tells us that, in borrowing the idea of utility from David Hume, his own contribution was to switch the interpretation from positive to normative. Many pages of his writings are devoted to lists of all the many factors that might make Pandora happy or sad. But the modern theory of rational decisions doesn't presume to tell you what your aims should be. Nor is it very specific about the means to be adopted to achieve whatever your ends may be. Indeed, since the modern theory can be regarded as a positive theory of behavior for an idealized world of rational agents, it needs to be explained why it has a normative role at all.

Pandora uses the theory of revealed preference normatively when she revises her attitudes to the world after discovering that her current attitudes would lead her to make choices in some situations that are inconsistent with the choices she would make in other situations.

A famous example arose when Leonard Savage was entertained to dinner by the French economist Maurice Allais. Allais asked Savage how he would choose in some difficult-to-assess situations (section 3.5). When Savage gave inconsistent answers, Allais triumphantly declared that even Savage didn't believe his own theory. Savage's response was to say that he had made a mistake. Now that he understood that his initial snap responses to Allais' questions had proved to generate inconsistencies, he would revise his planned choices until they became consistent.

One doesn't need to dine with Nobel laureates in Paris to encounter situations in which people use their rationality in revising snap judgments. I sometimes ask finance experts whether they prefer 96×69 dollars to 87×78 dollars. If given no time to think, most say the former. But when it is pointed out that $96 \times 69 = 6,624$ and $87 \times 78 = 6,786$, they always change their minds. An anecdote from Amos Tversky (2003) makes a similar point. In a laboratory experiment, many of his subjects made intransitive choices. When this was pointed out, a common response was to claim that his records were mistaken—the implication being that they wouldn't have made intransitive choices if they had realized they were doing so.

Positive applications. Rational decision theory isn't very good at predicting the choice behavior of inexperienced or unmotivated people. For example, laboratory experiments show that most people choose option (A) in problem 1 below and option (B) in problem 2, although the two problems differ only in how they are framed. Their choices are therefore unstable to irrelevant framing effects.

Problem 1. You are given $200 and then must choose between:

  (A) $50 extra for sure;

  (B) $200 extra with probability 25%.

Problem 2. You are given $400 and then must choose between:

  (A) $150 less for sure;

  (B) $200 less with probability 75%.

It is for this kind of reason that professors of marketing teach their students to laugh at economic consumer theory, which assumes that shoppers behave rationally when buying spaghetti or toothpaste.

However, there is a good deal of evidence that adequately motivated people sometimes can and do learn to choose rationally if they are put in similar situations repeatedly and provided with meaningful feedback on the outcome of their previous decisions (Binmore 2007a). Sometimes the learning is conscious, as when insurance companies systematically seek to maximize their long-term average profit. But most learning is unconscious. Like the squirrels who always find a way to get round the increasingly elaborate devices with which I have tried to defend my bird feeder, people mostly learn by trial-and-error. However, it is necessary to face up to the fact that the laboratory evidence suggests that humans find trial-and-error learning especially difficult when the feedback from our choices is confused by chance events, as will be the case in most of this book.

Fortunately, we don't just learn by trial-and-error. We also learn from books. Just as it is easier to predict how educated kids will do arithmetic, so the spread of decision theory into our universities and business schools will eventually make it easier to predict how decisions get made in economic life. If Pandora knows that $96 \times 69 = 6,624$ and $87 \times 78 = 6,786$, she won't make the mistake of choosing 96×69 dollars over 87×78 dollars. Once Allais had taught Savage that his choice behavior was inconsistent, Savage changed his mind about how to choose.

In brief, rational decision theory is only a useful positive tool when the conditions are favorable. Economists sometimes manage to convince themselves that the theory always applies to everything, but such enthusiasm succeeds only in providing ammunition for skeptics looking for an excuse to junk the theory altogether.

2

Game Theory

2.1 Introduction

Game theory is perhaps the most important arena for the application of rational decision theory. It is also a breeding ground for innumerable fallacies and paradoxes. However, this book isn't the place to learn the subject, because I plan to say only enough to allow me to use a few examples here and there. My book *Playing for Real* is a fairly comprehensive introduction that isn't mathematically demanding (Binmore 2007b).

2.2 What Is a Game?

A game arises when several players have to make decisions in a situation in which the outcome for each player is partly determined by the choices made by the other players. The states of the world that appear in Pandora's decision problem must therefore include the end products of the reasoning processes of her opponents in the game. Sometimes people speak of *strategic* uncertainty in such a situation.

A one-person decision problem in which Pandora has no opponents is sometimes called a one-player game, or a game against nature. The traditional aim of a game-theoretic analysis is to deduce how players will choose when playing a game against each other from how they each individually play games against nature. The latter information is assumed to be sufficiently complete that it can be summarized in terms of utilities that are called payoffs when they appear in the context of a game.

Critics commonly assume that a payoff in a game must be measured in dollars, and so game theorists get included in the class of mean-minded, money-grubbing misfits who supposedly think that everybody is as selfish as they are themselves. However, game theory is no less a daughter of the theory of revealed preference than any other branch of decision theory. It is therefore entirely neutral about what motivates the players, provided only that their choice behavior is consistent.

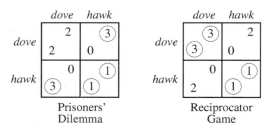

Figure 2.1. Paradox of rationality?

Nash equilibrium. Alice and Bob are using a Nash equilibrium in a two-player game if both choose a strategy that is a best reply to the strategy choice of the other (Nash 1951). There are two reasons why game theorists care about Nash equilibria.

The first reason is that a game theory book can't authoritatively point to a pair of strategies (a, b) as the rational solution of a game unless it is a Nash equilibrium. Suppose, for example, that b weren't a best reply to a. Bob would then reason that if Alice follows the book's advice and plays a, then he would do better not to play b. But a book can't be authoritative on what is rational if rational people don't play as it predicts (Binmore 2007b, p. 17).

Since this book is about making rational decisions, it is the rationalist defense of Nash equilibria that will always be relevant when game theory examples are mentioned. However, a second reason for caring about Nash equilibria is that they can be used to characterize the end product of an interactive evolutionary process (section 1.6). The idea is very simple. If the payoffs in a game correspond to how fit the players are, then adjustment processes that favor the more fit at the expense of the less fit will stop working when we get to a Nash equilibrium, because all the survivors will then be as fit as it is possible to be in the circumstances.

2.3   Paradox of Rationality?

The Prisoners' Dilemma is considered here as an example of how the payoffs in a game are determined by an appeal to the theory of revealed preference.

In the payoff table of figure 2.1, we assume that Alice must choose one of the rows labeled *dove* or *hawk*. Bob must simultaneously choose one of the columns labeled *dove* or *hawk*. The payoff that Alice receives is written in the southwest of each cell of the payoff table. Bob's payoff is written in the northeast of each cell.

The Prisoners' Dilemma is famous because it was once widely thought that it embodies the essence of the problem of human cooperation. It therefore seemed paradoxical that a rational analysis implies that both players will play *hawk* and so obtain a payoff of 1, when they could both play *dove* for a payoff of 2. Numerous fallacies purporting to show that it is rational to cooperate in the Prisoners' Dilemma were therefore invented (Binmore 1994, chapter 3).

Game theorists think it plain wrong to claim that the Prisoners' Dilemma is an appropriate setting within which to study the problem of human cooperation. On the contrary, it represents a situation in which the dice are as loaded against the emergence of cooperation as they could possibly be. If the game of life played by the human species resembled the Prisoners' Dilemma, we wouldn't have evolved as social animals!

Nor is there any paradox of rationality. Rational players don't cooperate in the Prisoners' Dilemma, because the conditions necessary for rational cooperation are absent. The following argument explains why game theorists are so emphatic in rejecting the idea that it might be rational to cooperate in the Prisoners' Dilemma.

Revealed preference in the Prisoners' Dilemma. So as not to beg any questions, we begin by asking where the payoff table that represents the players' preferences in the Prisoners' Dilemma comes from. The official answer is that we discover the players' preferences by observing the choices they make (or would make) when solving one-person decision problems.

What would Alice choose in the one-person decision problem she would face if she knew in advance that Bob is sure to play *dove* in the Prisoners' Dilemma? The circle in the southwest corner of the bottom-left cell of the payoff table indicates that we are given the information that she would then choose *hawk*. Similarly, the circle in the southwest corner of the bottom-right cell indicates that we are given the information that she would choose *hawk* if she knew in advance that Bob is sure to play *hawk*.

Writing a larger payoff for Alice in the bottom-left cell of the payoff table than in the top-left cell is just another way of registering that she would choose *hawk* if she knew that Bob were going to choose *dove*. Writing a larger payoff in the bottom-right cell is just another way of registering that she would choose *hawk* if she knew that Bob were going to choose *hawk*. If we want to retain a sense of paradox, we can also add the information that Alice would choose (*dove, dove*) over (*hawk, hawk*) if offered the opportunity. We must then ensure that Alice's payoff in the top-left cell exceeds her payoff in the bottom-right cell.

The circled payoffs are said to indicate Alice's best replies to Bob's strategies. This language invites the Causal Utility Fallacy, but we must remember that Alice doesn't choose *hawk* because she then gets a larger payoff. Alice assigns a larger payoff to (*hawk, dove*) than to (*dove, dove*) because she would choose the former if given the choice.

Game theory is needed because Alice has to choose her strategy in the Prisoners' Dilemma without knowing in advance what strategy Bob is going to choose. To predict what she will then do, we need to assume that she is sufficiently rational that the choices she makes in the game are consistent with the choices she would make when solving the simple one-person decision problems we have considered.

The trivial consistency requirement we require is a simplified variant of the sure-thing principle that we shall meet in a later chapter (section 7.2). I call it the umbrella principle to honor Reinhard Selten, who is a famous game theorist with an even more famous golfing umbrella. He always carries it on rainy days, and he always carries it on sunny days. But will he carry it tomorrow? We don't know whether tomorrow will be rainy or sunny, but the umbrella principle says that we don't need to know, because we can count on his carrying the umbrella anyway.

In the Prisoners' Dilemma, our data says that Alice will choose *hawk* if she learns that Bob is to play *dove*, and that she will also choose *hawk* if she learns that he is to play *hawk*. She thereby reveals that her choice doesn't depend on what she knows about Bob's choice. If her behavior is consistent with the umbrella principle, she will therefore play *hawk* whatever she guesses Bob's choice will be.

In game theory, *hawk* is said to be a strongly dominant strategy for Alice. After circling best replies in a payoff table, it is easy to spot a strongly dominant strategy because only her payoffs in the corresponding row will be circled.

Collective rationality?   The Prisoners' Dilemma teaches the important lesson that rationality need not be good for a society. For example, everybody in a society of rational individuals can be made worse off if certain pieces of information become common knowledge. However, the response that we should therefore junk the orthodox theory in favor of some notion of collective rationality makes no sense. One might as well propose abandoning arithmetic because two loaves and seven fishes won't feed a multitude.

Transparent dispositions?   A common objection to the preceding analysis of the Prisoners' Dilemma denies its premises. People say that Alice *wouldn't* choose *hawk* if she knew that Bob were going to choose *dove*.

For example, Alice might choose *dove* if she knew that Bob were going to choose *dove* because she has a reciprocating disposition. If it is transparent to both players that they both have a reciprocating disposition, then we are faced with a game like the Reciprocator Game of figure 2.1. No strategy is now dominant. Instead, both the strategy pairs (*dove, dove*) and (*hawk, hawk*) are Nash equilibria, because they correspond to cells in which both payoffs are circled, which implies that both players are simultaneously making best replies to each other.

But the fact that (*dove, dove*) is a Nash equilibrium in the Reciprocator Game doesn't imply that we have found a reason why it is rational to cooperate in the Prisoners' Dilemma. We have only found a reason why it might be rational to cooperate in the Reciprocator Game.

Whether Alice would or wouldn't choose *hawk* if she knew that Bob were going to choose *dove* is an empirical issue that is irrelevant to what is rational in the Prisoners' Dilemma. If Alice wouldn't choose *hawk*, she wouldn't be playing the Prisoners' Dilemma, but it would still be rational for her to play *hawk* if she were to play the Prisoners' Dilemma.

2.3.1 Anything Goes?

Critics sometimes reject rational decision theory on the grounds that it is just a bunch of tautologies, which can be ignored because they fail to exclude any outcome whatever of the game being played. In making this criticism, they miss the point that rational decision theory is about means rather than ends. This isn't to say that the players' ends are unimportant; only that it isn't part of rational decision theory to determine them.

Determining a player's ends takes place outside the theory. For example, in the usual story that accompanies the Prisoners' Dilemma, Alice and Bob are Chicago gangsters who are offered incentives by the District Attorney to fink on each other. Depending on who does or doesn't fink, Alice and Bob will spend more or less time in jail. But rational choice theory can't tell them how much they ought to value their honor as thieves in terms of years in jail, because this is an empirical question.

Empirical questions are settled by looking at the data. Since Chicago gangsters were apparently unscrupulous in seeking to avoid being jailed, we are led to model the problem faced by Alice and Bob in Chicago as the Prisoners' Dilemma. Rational decision theory then says that the solution of their problem is for both to fink on the other by playing *hawk*. If Alice and Bob had been Mother Theresa and St Francis of Assisi, we would have been led to another game with a different solution.

2.4 Newcomb's Problem

Rational decision theory is sometimes criticized because it is supposedly unable to deal with Newcomb's problem. Rival decision theories have even been invented to accommodate the difficulties it is thought to create. It is relevant here because David Lewis (1979) argued that the Prisoners' Dilemma reduces to two back-to-back Newcomb problems.

Newcomb's problem involves two boxes that may have money inside. Pandora is free to take either the first box or both boxes. If she cares only for money, what choice should she make? This seems an easy problem. If *dove* represents taking only the first box and *hawk* represents taking both boxes, then Pandora should choose *hawk*, because this choice always results in her getting at least as much money as *dove*. That is to say, *hawk* dominates *dove*.

However, there is a catch. It is certain that there is one dollar bill in the second box. The first box may contain nothing or it may contain two dollar bills. The decision about whether to put money in the first box is made by Quentin, who knows Pandora so well that he can always make a perfect prediction of what she will do, whether or not she behaves rationally. Like Pandora, he has two choices, *dove* and *hawk*. His dove-like choice is to put two dollar bills in the first box. His hawkish choice is to put nothing in the first box. His motivation is to catch Pandora out. He therefore plays *dove* if and only if he predicts that Pandora will choose *dove*. He plays *hawk* if and only if he predicts that Pandora will choose *hawk*.

Pandora's choice of *hawk* now doesn't look so good. If she chooses *hawk*, Quentin predicts her choice and puts nothing in the first box, so that Pandora gets only the single dollar in the second box. If Pandora chooses *dove*, Quentin predicts her choice and puts two dollars in the first box. Pandora then gets two dollars, but is left regretting the dollar in the second box that she failed to pick.

Robert Nozick (1969) argues that Newcomb's problem shows that maximizing your payoff can be consistent with using a strictly dominated strategy, but this obviously can't be right. I think that we are led to this contradiction because the premises of the Newcomb problem are contradictory. Binmore (1994, p. 242) shows that it is simply impossible to write down a game in which

1. Pandora has a genuine choice;
2. Quentin predicts before Pandora chooses;
3. Quentin's prediction is always accurate whatever Pandora may choose.

	dove	hawk
dove	$2	$0
hawk	$3	$1

Lewis

	yes	no
dove	$2	$0
hawk	$1	$3

Ferejohn

Figure 2.2. Newcomb's problem? Lewis reduces Pandora's problem to that of the row player in the Prisoners' Dilemma. In so doing, he fails to capture the assumption in Newcomb's problem that Quentin will predict her choice whether it is rational or not. Ferejohn makes the states of the world correspond to whether Quentin predicts Pandora's choice correctly or not. In so doing, he violates Aesop's principle.

For example, David Lewis's claim that the Prisoners' Dilemma is two back-to-back Newcomb problems fails to take into account that Quentin must predict Pandora's choice even if she were to choose irrationally. However, going over this old ground would take us too far afield, and so I shall simply draw attention to two attempts to solve Newcomb's problem that fail to honor Aesop's principle.

Richard Jeffrey's (1965) decision theory allows Pandora to count her own choice of action as evidence of the prediction that Quentin will make. In so doing, she fails to separate what is going on in her set A of actions from her beliefs in the set B of states of the world.

According to Steve Brams (1975), John Ferejohn suggests modeling the problem as in figure 2.2. In this model, the states of the world represent Quentin's success in predicting Pandora's choice. They are therefore labeled *yes* and *no*. If the probability that Quentin guesses correctly is sufficiently large, Pandora will then choose *dove*. However, Ferejohn's model violates Aesop's principle because the definitions of the states in B depend on what action is chosen in A.

2.5 Extensive Form of a Game

The payoff tables of figure 2.1 are strategic forms, in which the players are envisaged as choosing their strategies simultaneously. When the order of moves is important, a game is represented as an extensive form.

Figure 2.3 shows two versions of an extensive form for the Prisoners' Dilemma. In both cases, the game is represented as a tree whose root corresponds to the opening move. Alice moves first in one case and Bob does so in the other. It doesn't matter who is treated as moving first, because the information set enclosing the second player's two possible moves indicates that he or she doesn't know whether the first player chose *dove* or *hawk* at the first move.

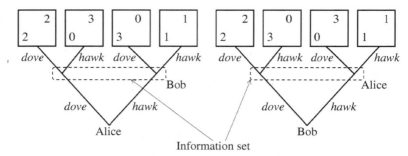

Figure 2.3. An extensive form. The figure shows two extensive forms for the Prisoners' Dilemma. In the left-hand case, Alice moves first. In the right-hand case, Bob moves first. It doesn't matter who moves first, because the information set enclosing the second player's two possible moves indicates that he or she doesn't know whether the first player chose *dove* or *hawk* at the first move.

2.5.1 Backward Induction

To analyze a game by backward induction, one assumes that Nash equilibria will be played, not only in the game as a whole, but in all of its subgames. Starting with the smallest subgames, one then replaces each subgame by a pair of payoffs that would be obtained by playing a Nash equilibrium in that subgame if it were reached.

The method is illustrated in figure 2.4 for the Prisoner's Dilemma of figure 2.1 played twice in succession. The repeated game has four subgames, consisting of four copies of the one-shot Prisoners' Dilemma that might be reached at the second stage of the repeated game depending on how Alice and Bob play at the first stage. The four copies differ because their payoffs take account of how much Alice and Bob gained at the first stage. For example, if the strategy pair (*dove, hawk*) is played at the first stage, then a payoff of 3 is added to each of Bob's payoffs at the second stage.

Since *hawk* dominates *dove* in each subgame, we replace each subgame by the payoff pair that is obtained when Alice and Bob both play *hawk* in the subgame. The result is the game on the right of figure 2.4, in which *hawk* again dominates *dove*. A backward induction analysis of the twice-repeated Prisoners' Dilemma therefore predicts that each player will always play *hawk*.

2.5.2 Possible Worlds

The chief reason for introducing games in extensive form is to draw attention to the importance of the idea of possible worlds. This idea was introduced by Leibniz and taken up in modern times by David Lewis

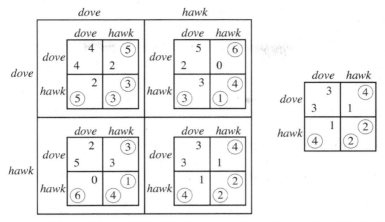

Figure 2.4. Playing the Prisoners' Dilemma twice. Alice and Bob's final pay-offs are the sum of the payoffs in the two games. There are four possible sub-games that can result from the players' strategy choices at the first stage of the repeated game. If the dominant strategy were played in each of these sub-games, the result would be the game on the right, in which it is again a dominant strategy to play *hawk*. Backward induction in the finitely repeated Prisoners' Dilemma therefore always calls for the play of *hawk*.

(1976), who entertained the delightful conceit that all possible worlds actually exist in some metaphysical sense.

Possible worlds matter a great deal in game theory because they help Pandora make sense of subjunctive conditionals of the form,

What would happen if I were to do that?

Such conditionals frequently express counterfactuals. In particular, if a rational analysis persuades Pandora to take action *a* rather than action *b*, then she won't take action *b*. But the reason that she takes action *a* is that she believes that if she were to take action *b*, she wouldn't get a better payoff.

In the twice-repeated Prisoners' Dilemma, there are four possible worlds that Alice and Bob need to consider. Each possible world corresponds to playing a copy of the Prisoners' Dilemma at the second stage after experiencing one of the four possible ways the Prisoners' Dilemma might be played at the first stage. The players need to predict what would happen if each of these possible worlds were to be reached.

In the case of the repeated Prisoners' Dilemma, the prediction is easy because *hawk* strongly dominates *dove* and it is therefore optimal to play *hawk* whatever a player might believe about the other player's plans. However, it is instructive for Alice if she observes Bob play *dove* at the first stage, because this tells her that Bob doesn't always play rationally.

So there is some chance that he will play irrationally in the future. In a game less simple than the twice-repeated Prisoners' Dilemma, this information might well lead her to seek to exploit Bob's perceived irrationality by deviating from Nash equilibrium play in some subgame—thereby subverting the logic of backward induction.

Bob Aumann (1995) argues that common knowledge of the players' rationality nevertheless implies that the backward induction path will be followed in finite games of perfect information. To pursue this controversy would take us way off course, but the issue will be mentioned again briefly in section 8.5 when discussing how knowledge should be interpreted in games.

3

Risk

3.1 Risk and Uncertainty

Economists attach a precise meaning to the words *risk* and *uncertainty*. Pandora makes a decision under risk if unambiguous probabilities can be assigned to the states of the world in her belief space B. Otherwise, she decides under uncertainty. The importance of distinguishing decision problems in which unambiguous probabilities are available from those in which they aren't was first brought to the attention of the world by Frank Knight (1921). For this reason, people often speak of *Knightian* uncertainty to emphasize that they are using the word in its technical sense.

The archetypal case of risk is playing roulette in a casino. The archetypal case of uncertainty is betting at the race track. Who knows what the probability is that *Punter's Folly* will win? Does it even make sense to attribute a probability to such a one-off occurrence?

Problems about when it makes sense to attach probabilities to events are the subject of an ongoing controversy in philosophy, but these problems will be put on the shelf until chapter 6. Our concern in the current chapter will be with issues that are nicely illustrated by the following paradox.

Zeckhauser's paradox. Some bullets are loaded into a revolver with six chambers, as illustrated in figure 3.1. The cylinder is then spun and the gun pointed at your head. Would you be prepared to pay more to get one bullet removed when only one bullet was loaded, or when four bullets were loaded? People usually say they would pay more in the first case, because they would then be buying their lives for certain.

I made this mistake myself when the paradox was first put to me, but I changed my mind when I learned that my decision was inconsistent with my preferring being alive to being dead and having more money rather than less. However, we shall have to wait for the reasons why until after the Von Neumann and Morgenstern theory of making decisions under risk has been examined.

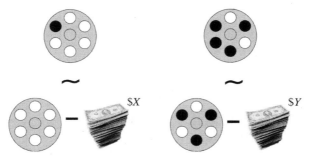

Figure 3.1. Zeckhauser's paradox. The diagram on the left shows that Pandora is indifferent between leaving one bullet in the revolver and paying X to get the bullet removed. The diagram on the right shows that Pandora is indifferent between leaving four bullets in the revolver and paying Y to get one bullet removed. Is Pandora's choice behavior consistent if $X > Y$?

3.2 Von Neumann and Morgenstern

John Von Neumann was an all-round genius. Inventing game theory was just a sideline for him. Critics of game theory choose to caricature him as the archetypal cold warrior—the original for Dr. Strangelove in the well-known movie. He was indeed a hawk in the cold war, but far from being a mad cyborg, he was a genial soul, whose parties were famous in Princeton for their relaxed atmosphere.

Von Neumann had proved the famous minimax theorem of game theory in 1928, but left the wider implications of the result unexplored until Oskar Morgenstern persuaded him that they should write a book they called *The Theory of Games and Economic Behavior* (Von Neumann and Morgenstern 1944), which put game theory on the map.

They were well into this enterprise when Morgenstern turned up at Von Neumann's house one day in the 1940s complaining that they didn't have a proper basis for the cardinal utilities they had been taking for granted in the book. Fortunately, Von Neumann didn't know or care that the reigning orthodoxy at the time was that cardinal utility scales are intrinsically nonsensical (section 1.7). So he invented a theory on the spot that measures how much Pandora wants something by the size of the risk she is willing to take to get it.

How does Von Neumann's theory work? Von Neumann's approach to utility is still sometimes regarded as an abstruse mathematical theory beyond the comprehension of ordinary folk. I think that this mistaken attitude is a hangover from the time when cardinal utility was controversial. After all, the old-time gurus who denounced cardinal utility as nonsensical were hardly likely to admit that they were wrong for a very simple reason.

Suppose that Pandora is wondering how much effort it is worth expending to get a date with Quentin tonight. How many utils should she assign to this outcome in her set C of consequences?

We first need to decide what utility scale to use. For this purpose, Pandora needs to pick two outcomes that are respectively better and worse than any other outcome she is likely to encounter. For example, the worst outcome might be that she has to spend the evening home alone. Her best outcome might be that she somehow gets a date with Johnny Depp. These two extremes will correspond to the boiling and freezing points of water used to calibrate a thermometer, except that the utility scale to be constructed will assign 0 utils to the worst outcome, and 1 util to the best outcome (rather than 32° and 212° as on the Fahrenheit scale).

Next consider a bunch of (free) lottery tickets for which the only prizes are either the best outcome or the worst outcome. When we offer Pandora lottery tickets with higher and higher probabilities of getting the best outcome as an alternative to a date with Quentin, she will eventually switch from saying *no* to saying *yes*. If the probability of the best outcome on the lottery ticket that makes her switch is 0.2, then Von Neumann's theory says that a date with Quentin is worth 0.2 utils to her.

Pandora needs to know that a date with Quentin is worth 0.2 utils in order to determine the expected value of her Von Neumann and Morgenstern utility function. But why should a rational agent maximize the long-run average of such a utility function and not something else? This question was first asked by Nicholas Bernoulli way back in the eighteenth century when he formulated the St Petersburg paradox.

3.3 The St Petersburg Paradox

In the heyday of the Czars, a casino in St Petersburg was supposedly willing to run any lottery whatever, provided that the management could set the price of a ticket to participate.[1]

In the lottery of figure 3.2, a fair coin is tossed until it shows heads for the first time. If the first head appears on the kth trial, a player wins $\$2^k$. How much should Pandora be willing to pay to participate in this lottery?

A player wins 2^k dollars in the St Petersburg lottery with probability $1/2^k$. For example, the probability of the sequence TTH, which wins

---

[1] The story is good, but the paradox probably got its name for the more prosaic reason that Daniel Bernoulli published it in the *Proceedings of the St Petersburg Academy* of 1738. The brothers Daniel and Nicholas are only two of a whole family of mathematical Bernoullis.

Prize	$2	$4	$8	$16	$
Coin sequence	H	TH	TTH	TTTH	TTT
Probability	1/2	1/4	1/8	1/16	

Figure 3.2. The St Petersburg paradox. How much should Pandora be willing to pay to win 2^k dollars with probability $1/2^k$?

8 dollars, is $1/2^3 = 1/8$. So a player expects to win

$$2/2 + 2^2/2^2 + 2^3/2^3 + \cdots = 1 + 1 + 1 + \cdots$$

dollars on average. The expected dollar value of the lottery is therefore infinite.

If it made sense for Pandora to choose whatever maximizes her dollar expectation, she would liquidate her entire assets and borrow whatever she could in order to buy a ticket for the lottery. But the probability is $\frac{7}{8}$ that Pandora will end up with no more than $8, and so she is unlikely to find the odds attractive.

The moral isn't that the policy of always choosing the lottery with the largest expectation in dollars is irrational. It isn't for us to tell Pandora what her aims should be. The St Petersburg story only casts doubt on the idea that no other policy can be rational. The same goes for any theory which claims that there is only one rational way to respond to risk. An adequate theory needs to recognize that the extent to which Pandora is willing to bear risk is as much a part of her preference profile as her attitude to getting a date with Quentin or Johnny.

Decreasing marginal utility. The St Petersburg paradox shows that maximizing your dollar expectation may not always be a good idea. It suggests that an agent in a risky situation might sometimes want to maximize the expectation of some utility function with decreasing marginal utility. This just means that each extra dollar is assigned a smaller number of utils than the dollar that went before. Paul Samuelson famously explained that this is why rich men take a cab when it rains but poor men get wet—the cab fare is worth more to you if you are poor than if you are rich. As Jeremy Bentham (1863) put it: "The quantity of happiness produced by a particle of wealth will be less and less at every particle."

If Pandora's utility for x dollars is $u(x)$, then her expected utility in the St Petersburg lottery becomes

$$u(2) \times \tfrac{1}{2} + u(4) \times \tfrac{1}{4} + u(8) \times \tfrac{1}{8} + \cdots.$$

The St Petersburg paradox assumes that $u(x) = x$, but Daniel Bernoulli suggested taking the utility function to be the logarithm of Pandora's dollar winnings. If we adopt his suggestion using logarithms to base 2,[2] then the expected utility of the St Petersburg lottery is reduced all the way from infinity to 2. It follows that Pandora would only be willing to pay $4 to participate in the St Petersburg lottery.

Daniel Bernoulli's conclusion accords with our intuitions about how ordinary people would behave in the mind experiment envisaged in the St Petersburg paradox. Nobody I know would be prepared to pay more than $10 to participate. But, although Daniel Bernoulli predates Bentham by many years, his proposal is Benthamite in character. We are left wondering what is special about the logarithm of income. Why not take Pandora's utility function to be the square root or some other function of income? For example, if $u(x) = 4\sqrt{x}$, then Pandora would be willing to pay $5.86 to participate in the St Petersburg lottery.

Why are we maximizing the long-run average of a utility function anyway? Why shouldn't Pandora be ultracautious and maximize her minimum possible income, as assumed by John Rawls (1972) in his famous *Theory of Justice*? Since her minimum possible prize in the St Petersburg lottery is $2, she would then be unwilling to pay more than this amount to participate.

The theory of revealed preference allows us to see our way around such questions. In particular, the Von Neumann and Morgenstern theory of decision under risk explains why Pandora will necessarily act as though maximizing the expected value of something if she honors their consistency requirements.

3.4 Expected Utility Theory

This section offers a simple version of Von Neumann and Morgenstern's theory.

Lotteries. In a decision problem $D : A \times B \rightarrow C$ the feasible set A of actions is a subset of the set $\aleph$ of all acts $\alpha : B \rightarrow C$ (section 1.2). In the Von Neumann and Morgenstern theory, each state of the world in Pandora's belief set B comes equipped with a precise probability. When making decisions, Pandora is assumed to pay no attention to any other features of a state.

In the Von Neumann and Morgenstern theory, Pandora's beliefs over the states in the set B and her preferences over the consequences in the

---

[2] Recall that $k = \log_2 2^k$, and so Pandora's expected utility is $\sum_{k=1}^{\infty} k \times (\frac{1}{2})^k = 2$.

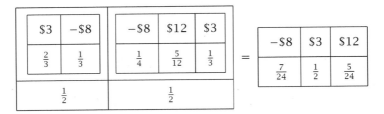

Figure 3.3. Reducing a compound lottery. For example, the total probability of losing \$8 in the compound lottery is $\frac{1}{2} \times \frac{1}{3} + \frac{1}{2} \times \frac{1}{4} = \frac{7}{24}$.

set C are separated right from the outset. The probabilities of the states in B are assumed to be exogenously determined, and so not open to any bias that Pandora's preferences might induce. Nor does the theory allow any bias to arise from the nature of Pandora's feasible set A, since a Von Neumann and Morgenstern utility function is derived from the preferences she reveals over the whole set $\aleph$ of acts.

Given that only the probabilities of states are to be counted as relevant, the first step in the derivation is to identify $\aleph$ with the set lott(C) of lotteries with prizes in C. For example, if C consists of possible money payments, then the lottery in which Pandora loses \$8 with probability $\frac{1}{4}$, wins \$12 with probability $\frac{5}{12}$, and wins \$3 with probability $\frac{1}{3}$ is represented by

$$L = \begin{array}{|c|c|c|} \hline -\$8 & \$12 & \$3 \\ \hline \frac{1}{4} & \frac{5}{12} & \frac{1}{3} \\ \hline \end{array}.$$

The expected dollar value of this lottery is

$$\mathcal{E}L = -8 \times \tfrac{1}{4} + 12 \times \tfrac{5}{12} + 3 \times \tfrac{1}{3} = 4.$$

One of the prizes in a raffle at an Irish county fair is sometimes a ticket for the Irish National Sweepstake. If you buy a raffle ticket, you are then participating in a compound lottery in which the prizes may themselves be lotteries. All the lotteries involved in a compound lottery are always assumed to be *independent* of each other (unless something is said to the contrary), and so it is easy to reduce a compound lottery to a simple lottery, as in figure 3.3.

3.4.1 Rationality Postulates

The following list of postulates are simplifications of those proposed by Von Neumann and Morgenstern (1944).[3] They require that the

---

[3] Von Neumann and Morgenstern's own exposition is best read as presented in the third edition (1953) of their book on pages 15–31.

preferences over lotteries that Pandora reveals in risky situations be consistent.

Postulate 1. Pandora prefers whichever of two win-or-lose lotteries offers the larger probability of winning.

Recall from section 3.2 that Pandora must identify outcomes that are respectively better and worse than any outcome she needs to consider in her decision problem. These outcomes will be called $\mathcal{W}$ and $\mathcal{L}$. A win-or-lose lottery is simply a lottery in which the only outcomes are $\mathcal{W}$ and $\mathcal{L}$.

Postulate 2. Pandora is always indifferent between each prize $\mathcal{P}$ and some lottery involving only $\mathcal{W}$ and $\mathcal{L}$.

The second postulate says that, for each prize $\mathcal{P}$ in Pandora's set of consequences, there is a probability q for which

$$\mathcal{P} \sim \begin{array}{|c|c|} \hline \mathcal{W} & \mathcal{L} \\ \hline q & 1-q \\ \hline \end{array}. \tag{3.1}$$

The postulate is sometimes called the Archimedean axiom for reasons that aren't worth remembering. It is needed to justify the method used to construct a Von Neumann and Morgenstern utility function $u : C \to \mathbb{R}$ in section 3.2. We simply make $u(\mathcal{P})$ equal to the probability q of (3.1).

However, it isn't enough to have constructed a utility function that represents Pandora's revealed preferences over the prizes in her consequence space C. For u to be a Von Neumann and Morgenstern utility function, Pandora must choose among the lotteries in lott(C) as though she were maximizing the expected utility $\mathcal{E}u(L)$ of the lottery L. Not only do we require that u represents Pandora's preferences over C, we therefore also require that $\mathcal{E}u$ represents Pandora's preferences over lott(C).

Figure 3.4 illustrates the two steps in the argument that justifies this conclusion. Each step requires a further postulate.

Postulate 3. Pandora doesn't care if a prize in a lottery is replaced by an independent prize if she is indifferent between the new prize and the prize it replaces.

Versions of postulate 3 are often called the independence axiom. This terminology serves as a reminder that it is implicitly assumed that if one of the prizes is itself a lottery, then this lottery must be independent of all other lotteries involved.

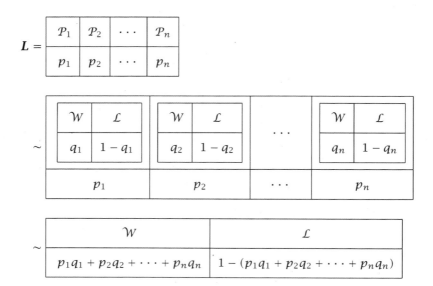

Figure 3.4.   Von Neumann and Morgenstern's argument. The first indifference is justified by postulate 3. The second indifference is justified by postulate 4.

The prizes available in the arbitrary lottery L of figure 3.4 are $\mathcal{P}_1, \mathcal{P}_2, \ldots, \mathcal{P}_n$. By postulate 2, Pandora is indifferent between each such prize $\mathcal{P}_k$ and some win-or-lose lottery in which she wins with probability q_k. Postulate 3 is then used to justify replacing each prize $\mathcal{P}_k$ by the corresponding win-or-lose lottery. We then need a final assumption to reduce the resulting compound lottery to a simple lottery.

Postulate 4.   Pandora only cares about the total probability with which she gets each prize in a compound lottery.

The total probability of $\mathcal{W}$ in figure 3.4 is $r = p_1 q_1 + p_2 q_2 + \cdots + p_n q_n$. So postulate 4 says that we can replace the compound lottery by the win-or-lose lottery on the final line, thereby justifying the second of the two steps the figure illustrates.

By postulate 1, Pandora prefers whichever of two lotteries like L in figure 3.4 has the larger value of $r = p_1 q_1 + p_2 q_2 + \cdots + p_n q_n$. She therefore acts as though seeking to maximize

$$r = p_1 q_1 + p_2 q_2 + \cdots + p_n q_n$$
$$= p_1 u(\mathcal{P}_1) + p_2 u(\mathcal{P}_2) + \cdots + p_n u(\mathcal{P}_n)$$
$$= \mathcal{E}u(L).$$

It follows that $\mathcal{E}u$ is a utility function that represents Pandora's preferences over lotteries. But this is what it means to say that u is a

Von Neumann and Morgenstern utility function for her preferences over prizes.

Standing Quentin up. The following example emphasizes the importance of the independence proviso in the vital postulate 3.

Suppose that Pandora is indifferent between a date with Quentin and the lottery **Q** in which a fair coin is tossed that results in her getting a date with Johnny Depp if it falls *heads* and a date with nobody at all if it falls *tails*. Now imagine that Pandora is offered a lottery **L** in which she gets a date with Johnny if a fair coin falls *tails* and a date with Quentin if it falls *heads*. Will she be ready to swap the unfortunate Quentin in this lottery for **Q**?

She most certainly would if we were talking about the *same* coin toss in each lottery, because she would then be guaranteed a date with Johnny whatever happens! Even if the coins were different but positively correlated she would still strictly prefer standing Quentin up.

3.5 Paradoxes from A to Z

This section examines Allais' paradox and Zeckhauser's paradox as examples of how the Von Neumann and Morgenstern theory works in practice.

Allais' paradox. Allais asked Savage to compare the lotteries **J** and **K** of figure 3.5, and then to compare the lotteries **L** and **M** (section 1.10). When Savage reported that he would choose **J** rather than **K**, and **M** rather than **L**, Allais triumphantly pointed out that the preferences he thereby revealed are inconsistent with the Von Neumann and Morgenstern theory.

To see that the preferences $K \prec J$ and $L \prec M$ violate the Von Neumann and Morgenstern postulates, it is only necessary to show that they can't be described by a Von Neumann and Morgenstern utility function.

The best and worst prizes available are $0 and $5m. We therefore look for a Von Neumann and Morgenstern utility function with $u(0) = 0$ and $u(5) = 1$. What can then be said about Savage's value for $x = u(1)$? Observe that

$$\mathcal{E}u(J) = u(0) \times 0.0 + u(1) \times 1.0 + u(5) \times 0.0 \quad = x,$$
$$\mathcal{E}u(K) = u(0) \times 0.01 + u(1) \times 0.89 + u(5) \times 0.10 = 0.89x + 0.10,$$
$$\mathcal{E}u(L) = u(0) \times 0.89 + u(1) \times 0.11 + u(5) \times 0.0 \quad = 0.11x,$$
$$\mathcal{E}u(M) = u(0) \times 0.90 + u(1) \times 0.0 + u(5) \times 0.10 \quad = 0.10.$$

Since $J \succ K$, we have that $x > 0.89x + 0.10$, and so $x > \frac{10}{11}$. Since $L \prec M$, we also have that $0.11x < 0.10$, and so $x < \frac{10}{11}$. But it can't be

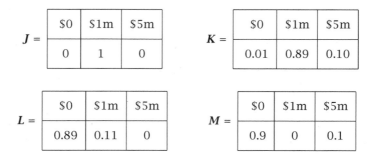

Figure 3.5. Allais' paradox. The prizes are given in millions of dollars to dramatize the situation.

simultaneously true that $x > \frac{10}{11}$ and $x < \frac{10}{11}$. So the preferences that Savage expressed can't be described with a Von Neumann and Morgenstern utility function.

Experiments are said to show that most laboratory subjects express the same preferences as Savage. Allais' paradox then provides ammunition for people who deny that the Von Neumann and Morgenstern theory is a good model of how real people behave in risky situations. My own view is that such critics are right to challenge the kind of positive application of the theory that is standard in economics, but wrong to think that Allais' paradox adds much to their case.

Most people presumably express the preference $J \succ K$ because J guarantees $1m for sure, whereas K carries the risk of getting nothing at all. On the other hand, when M is compared with L, the risk of ending up with nothing at all can't be avoided. On the contrary, the risk of this final outcome is high in both cases. But if the probability 0.89 in L is rounded up to 0.90 and 0.11 is rounded down to 0.10, then someone who understands what is going on will prefer M to the new L. If the new L is thought to be essentially the same as the old L, one then has a reason for expressing the preference $M \succ L$. In other words, the version of the Allais' paradox presented here can be explained in terms of rounding errors.

Zeckhauser's paradox. Suppose, as in figure 3.1, that X is the most that Pandora is willing to pay to get one bullet removed from a gun containing one bullet, and that Y is the most that she is willing to pay to get one bullet removed from a gun containing four bullets (section 3.1). Let $\mathcal{L}$ mean death, and $\mathcal{W}$ mean being alive after paying nothing. Let C mean being alive after paying X, and $\mathcal{D}$ mean being alive after paying Y.

The aim is to assign Von Neumann and Morgenstern utilities to the consequences C and $\mathcal{D}$. If we find that $u(\mathcal{D}) < u(C)$, then it follows that $\mathcal{D} \prec C$, and so it must be the case that $X \prec Y$ because Pandora is assumed to prefer more money to less.

We shall follow our practice so far of making $u(\mathcal{L}) = 0$ and $u(\mathcal{W}) = 1$ (section 3.2). Sometimes people complain that death is so terrible an outcome that it should be assigned a utility of $-\infty$. But even if this were allowed in the Von Neumann and Morgenstern theory, it would be completely unrealistic. If $u(\mathcal{L}) = -\infty$, then Pandora wouldn't be willing to do anything whatever that involved the slightest risk of dying. She wouldn't even go for a walk in the park in case a meteor fell on her head. In fact, estimates derived from driving behavior suggest that ordinary people value their lives at something less than ten million dollars.

Returning to Zeckhauser's paradox, we are given that Pandora is indifferent between C and the lottery in which she gets $\mathcal{L}$ with probability $\frac{1}{6}$ and $\mathcal{W}$ with probability $\frac{5}{6}$. Thus,

$$u(C) = \tfrac{1}{6}u(\mathcal{L}) + \tfrac{5}{6}u(\mathcal{W}) = \tfrac{5}{6}.$$

Similarly, Pandora is indifferent between the lottery in which she gets $\mathcal{L}$ and $\mathcal{D}$ each with probability $\frac{1}{2}$, and the lottery in which she gets $\mathcal{L}$ with probability $\frac{2}{3}$ and $\mathcal{W}$ with probability $\frac{1}{3}$. Thus,

$$\tfrac{1}{2}u(\mathcal{L}) + \tfrac{1}{2}u(\mathcal{D}) = \tfrac{2}{3}u(\mathcal{L}) + \tfrac{1}{3}u(\mathcal{W}),$$
$$u(\mathcal{D}) = \tfrac{2}{3}.$$

So $\mathcal{D} \prec C$, and thus Pandora will necessarily be ready to pay less to get one bullet removed when only one bullet was loaded than when four bullets were loaded.

After seeing the calculation, the result begins to seem more plausible. Would I be willing to pay more to get a bullet removed from a six-shooter containing *one* bullet than to get a bullet removed from a six-shooter containing *six* bullets? Definitely not! But getting a bullet removed when there are six bullets isn't so different from getting a bullet removed when there are five bullets, which isn't so different from getting a bullet removed when there are four bullets. *How* different is the difference between each of these cases? Appealing to our gut feelings doesn't get us very far when such questions are asked. We need to calculate.[4]

---

[4]Kahneman and Tversky (1979) think the example misleading on the grounds that matters are confused by the question of whether money has value for you after you are dead. However, the result would remain the same even if one were to distinguish different types of death depending on how much money you left for your heirs. The only necessary assumption would be that the Von Neumann and Morgenstern utilities of these different types of death not be too far apart.

3.6 Utility Scales

We have seen that there is always an infinite number of possible utility functions that represent any consistent preference relation. If U is an ordinal utility function, then any strictly increasing transformation of U is also an ordinal utility function that represents the same preference relation (section 1.7). However, a Von Neumann and Morgenstern utility function u is cardinal. Given any two Von Neumann and Morgenstern utility functions u_1 and u_2 that represent the same preferences over lott(C), we can write

$$u_2 = Au_1 + B, \tag{3.2}$$

where $A > 0$ and B are constants.

3.6.1 Analogy with Temperature

Von Neumann and Morgenstern emphasize the analogy between their kind of utility scale and temperature scales. In both cases, we are free to assign the zero and the unit on the scale in any way we find convenient. We therefore assign $0°$ to the freezing point of water on the Celsius scale and $100°$ to the boiling point. The reason that it was once convenient to assign $32°$ to the freezing point of water on the Fahrenheit scale and $212°$ to the boiling point of water are nowadays forgotten. However, it doesn't matter very much that the Fahrenheit scale seems arbitrary, because we can pass back and forward between the two scales using the affine transformation $f = \frac{9}{5}c + 32$.

In a similar way, we don't need to use a utility scale that assigns a utility of 0 to $\mathcal{L}$ and 1 to $\mathcal{W}$. We can assign them any utilities we find convenient, provided that the utility we assign to $\mathcal{W}$ exceeds the utility we assign to $\mathcal{L}$. However, once we have determined the zero and the unit on a Von Neumann and Morgenstern utility scale, we have exhausted all our room for maneuver.

Justifying the analogy. What justifies equation (3.2)? It is obvious that maximizing the expected utility of u_1 is equivalent to maximizing the expected utility of $u_2 = Au_1 + B$. It is not so obvious that if u_1 and u_2 are alternative Von Neumann and Morgenstern utility functions for a preference relation $\preceq$ defined on lott(C), then they must be linked as in (3.2).

To see why, begin by choosing constants $A_i > 0$ and B_i to make the Von Neumann and Morgenstern utility functions $v_i = A_i u_i + B_i$ satisfy $v_i(\mathcal{L}) = 0$ and $v_i(\mathcal{W}) = 1$. Recall that postulate 2 implies that, for any prize $\mathcal{P}$ in C, there is a probability q for which

$$\mathcal{P} \sim \begin{array}{|c|c|} \hline \mathcal{W} & \mathcal{L} \\ \hline q & 1-q \\ \hline \end{array} .$$

This observation allows us to deduce that $v_1(\mathcal{P}) = v_2(\mathcal{P})$ because both sides are equal to $(1-q)v_i(\mathcal{L}) + qv_i(\mathcal{W}) = q$. Then rewrite v_i in terms of u_i, and solve for u_2 in terms of u_1.

The domain of a Von Neumann and Morgenstern utility function. The domain of a Von Neumann and Morgenstern utility function $u : C \to \mathbb{R}$ is Pandora's set C of consequences. The domain of its expected value $\mathcal{E}u : \text{lott}(C) \to \mathbb{R}$ is the set $\text{lott}(C)$ of lotteries with prizes in C.

Confusion can arise because some authors say that $\mathcal{E}u$ is the Von Neumann and Morgenstern utility function rather than u. The difference is large because $\mathcal{E}u$ has only ordinal status on $\text{lott}(C)$, whereas u has cardinal status on C. It is hard to believe in retrospect, but this misunderstanding once created a major controversy in the utilitarianism literature (Elster and Roemer 1992).

Von Neumann and Morgenstern utility functions are sometimes said to be linear for similar reasons. There need be nothing linear about $u : C \to \mathbb{R}$, but there is a sense that we need not worry about in which $\mathcal{E}u : \text{lott}(C) \to \mathbb{R}$ always defines a linear mapping from a set of probability vectors to the real numbers.

3.6.2 Isolating Decision Problems

The lectures that gave rise to this book were given in honor of William (Terence) Gorman. This section uses his idea that the satisfaction people find in commodities can sometimes be regarded as deriving from separable characteristics (Gorman and Myles 1988). I hope he would also have liked my pursuing the analogy of utility with temperature even to the extent of contemplating the possibility of an absolute zero on a Von Neumann and Morgenstern utility scale.

Separating preferences. Pandora may like oranges for two reasons: they are healthy and they taste nice. Her preferences can then be separated into a health dimension and a taste dimension. Pandora needs to be able to separate her preferences in this kind of way, because she would otherwise never be able to squeeze a decision problem into a small-world format. Even deciding between a vanilla and a strawberry ice cream would become problematic if she always had to take account of the totality of

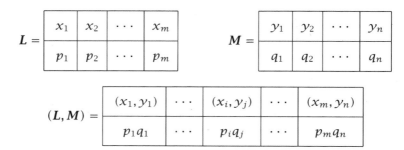

Figure 3.6. Evaluating lotteries separately. The lottery (L, M)
is obtained by combining the separate lotteries L and M.

her environment when doing so. The ability to isolate only the preference dimensions that matter in a particular problem is therefore of great
importance. But how can this be done?

The first step in addressing the separation question is to write $C = C_1 \times C_2$, so that each consequence $c = (c_1, c_2)$ in C can be evaluated in
terms of two characteristics c_1 and c_2. If Pandora evaluates these two
dimensions separately, then her preferences on one dimension should
be unaffected by how well she is doing on the other dimension. When this
requirement is expressed in terms of lotteries, the result is surprisingly
strong (Keeney and Raiffa 1975).

Figure 3.6 shows two lotteries L and M with prizes in C_1 and C_2 respectively. The lottery (L, M) (with prizes in $C = C_1 \times C_2$) is one particular way of combining L and M. We shall say that a preference relation
$\preceq$ defined on $\text{lott}(C)$ evaluates $\text{lott}(C_1)$ and $\text{lott}(C_2)$ separately if it is
always true that

(1) $(L, M) \prec (L, M')$ implies $(L', M) \preceq (L', M')$;

(2) $(L, M) \prec (L', M)$ implies $(L, M') \preceq (L', M')$.

Separating utilities. If $\preceq$ satisfies the Von Neumann and Morgenstern
postulates, then we can represent it with a Von Neumann and Morgen-
stern utility function $u : C \to \mathbb{R}$ satisfying $u(\mathcal{L}) = 0$ and $u(\mathcal{W}) = 1$,
where $\mathcal{W} = (\mathcal{W}_1, \mathcal{W}_2)$ and $\mathcal{L} = (\mathcal{L}_1, \mathcal{L}_2)$ are the best and worst outcomes
in C. We shall also need to consider $A = u(\mathcal{W}_1, \mathcal{L}_2)$ and $B = u(\mathcal{L}_1, \mathcal{W}_2)$.

If $\preceq$ evaluates $\text{lott}(C_1)$ and $\text{lott}(C_2)$ separately, section 10.1 shows that

$$u = u_1 u_2 + A u_1 (1 - u_2) + B u_2 (1 - u_1), \qquad (3.3)$$

where the functions $u_1 : C_1 \to \mathbb{R}$ and $u_2 : C_2 \to \mathbb{R}$ can be regarded as
Von Neumann and Morgenstern utility functions on C_1 and C_2, with
$u_1(\mathcal{L}_1) = u_2(\mathcal{L}_2) = 0$ and $u_1(\mathcal{W}_1) = u_2(\mathcal{W}_2) = 1$.

Three cases need to be distinguished: $A + B < 1$, $A + B = 1$, and $A + B > 1$. Rationality can't decide between these cases. Pandora reveals which case applies to her when she chooses between the lottery L, in which she has an equal chance of getting $(\mathcal{L}_1, \mathcal{L}_2)$ or $(\mathcal{W}_1, \mathcal{W}_2)$, and the lottery M, in which she has an equal chance of getting $(\mathcal{L}_1, \mathcal{W}_2)$ or $(\mathcal{W}_1, \mathcal{L}_2)$. Case 1 applies if she always chooses L. Case 3 applies if she always chooses M. Case 2 applies if she is indifferent.

Absolute zero? Where the matter is considered at all, the rational choice literature commonly makes assumptions that instantiate case 2 (Fishburn and Rubinstein 1982). More often, as in section 2.5.1, it is simply taken for granted that Pandora behaves as though seeking to maximize the (discounted) sum of utilities derived from separate consequence spaces. I have used case 1 in defending the Nash bargaining solution, but this is quite unusual (Binmore 1984).

Without denying that all three cases may arise in different applications, I think that case 1 is the best candidate for a default assumption. To see why, suppose that C_1 is the set of consequences that appears in Pandora's decision problem, and C_2 takes account of everything else. To deny case 1 is then to deny that the worst outcome $\mathcal{L}_2$ in C_2 is so bad that its realization makes Pandora regard the distinctions between the consequences of C_1 as irrelevant. In Zeckhauser's paradox, Pandora's death eliminated her interest in money (section 3.1), but the bad event need not be so drastic. For example, if Pandora learns that Quentin is unfaithful, she may temporarily cease to care what lipstick she wears.

The point here is that if $A = u(\mathcal{W}_1, \mathcal{L}_2) = 0$, then $A + B = B \leqslant 1$. If it is also true that $B = u(\mathcal{L}_1, \mathcal{W}_2) = 1$, then $u(c_1, c_2) = u_2(c_2)$, so that Pandora doesn't ever care what consequence is realized in C_1.[5] If this extreme possibility is eliminated, then $B < 1$ and so we are left with case 1.

When case 1 applies, we have a reason for identifying an absolute zero on Pandora's Von Neumann and Morgenstern utility scale. Since no temperature can fall below $-273.15\ °C$, scientists make this the zero temperature on the Kelvin scale, and say that $0\ °K$ is absolute zero. Similarly, no utility can fall below whatever Pandora gets when $\mathcal{L}_2$ occurs, and so it is natural to use a Von Neumann and Morgenstern utility scale whose zero corresponds to this outcome.

---

[5] When $\mathcal{L}_2$ is hell and $\mathcal{W}_2$ is heaven, who cares about the joys and sorrows of our earthly lives?

3.7 Attitudes to Risk

Pandora must choose between:

1. an equal chance of getting either $4 or $36;
2. an equal chance of getting $9 or $25.

Pandora's Von Neumann and Morgenstern utility for x is $\sqrt{x}$, and so her expected utility for both lotteries is 4 utils, but she tells us that she plans to choose the second lottery because it has a smaller variance, and she is averse to taking risks.[6]

Pandora's story is irrational, because a rational person is necessarily indifferent between two lotteries if their expected utilities are the same. We aren't thereby denying that Pandora might be risk averse. Her risk aversion is built into the shape of her utility function (figure 3.9). In general, the Von Neumann and Morgenstern utility functions of risk-averse people are concave, and the Von Neumann and Morgenstern utility functions of risk-loving people are convex (figure 3.7).

Note that the degree of risk aversion a person reveals is a matter of personal preference in the Von Neumann and Morgenstern theory. Just as Pandora may or may not prefer Mozart to Wagner, so she may or may not prefer to spend $500 on insuring her house. Some philosophers insist to the contrary that rationality implies prudence, and it is prudent for Pandora to be risk averse. But such considerations go beyond the consistency requirements of the Von Neumann and Morgenstern theory.

Concavity.   Concave functions have many pleasant properties. In particular, the slope of a concave function always decreases. The slope $u'(x)$ is Pandora's marginal utility at x.[7] So a risk-averse person has decreasing marginal utility (section 3.3).

A function can be simultaneously concave and convex. Its graph is then a straight line. Mathematicians say that such a function is affine. If Pandora has an affine utility function, she is said to be risk neutral.

Figure 3.8 shows an affine utility function alongside a utility function that is neither concave nor convex. Friedman and Savage (1948) suggest that we can explain the behavior of people who insure their houses against fire but take their vacations in Las Vegas by attributing such a utility function to them. They are risk loving for small amounts and so

---

[6] The variance of a lottery measures how far the money prizes deviate from their mean value.

[7] When working with a continuous variable x, we assume that the units δx in which x is measured are so small that the extra utility Pandora gains from one extra unit of x is approximately $u'(x)\delta x$.

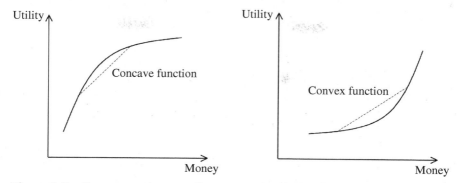

Figure 3.7. Concave and convex functions. Chords drawn to a concave function lie on or below the graph. Chords drawn to a convex function lie on or above the graph.

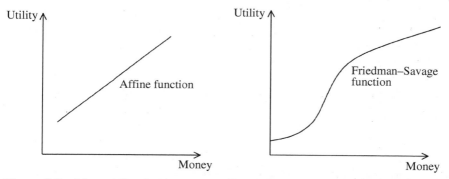

Figure 3.8. More utility functions. An affine function has a straight-line graph. The Friedman–Savage function starts out being convex and then switches to being concave.

bet in casinos. They are risk averse for large amounts and so also take out insurance.

Risk aversion. Why are people with concave utility functions risk averse? The formal definition says that Pandora is risk averse if she always prefers the dollar average of any set of money prizes for certain to participating in a lottery in which she has an equal chance of ending up with each prize. In mathematics:

$$u(\mathcal{E}L) \geqslant \mathcal{E}u(L), \tag{3.4}$$

where L is a lottery with money prizes and u is Pandora's Von Neumann and Morgenstern utility function. Figure 3.9 explains why this criterion holds if and only if u is concave.

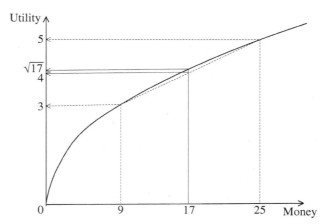

Figure 3.9. Risk aversion. Pandora's Von Neumann and Morgenstern utility for x is $\sqrt{x}$. The expected utility of the lottery L in which Pandora has an equal chance of getting $9 or $25 is $4 = (\sqrt{9} + \sqrt{25})/2$. Pandora therefore prefers getting $17 = (9 + 25)/2$ dollars for sure to L because $\sqrt{17} > \sqrt{16}$. Her utility function u therefore satisfies the requirement $u(\mathcal{E}L) \geqslant \mathcal{E}u(L)$ for risk aversion. The same argument goes through for any concave utility function u.

It is often useful in applications to know to what degree a person is risk averse. Economists use various coefficients of risk aversion for this purpose, which are explained very clearly by Hirshleifer and Riley (1992).

Maximin criterion. John Rawls (1972) argues that the maximin criterion corresponds to the case of extreme risk aversion, but this view is incompatible with the Von Neumann and Morgenstern theory.[8]

The maximin criterion says that Pandora should ignore everything about a lottery except the smallest prize that it offers with positive probability. She should then choose the lottery in her feasible set whose smallest prize is largest. For example, J should be chosen over K in figure 3.5 because J guarantees $1m, whereas there is a positive probability of getting nothing at all from K.

Figure 3.10 shows a sequence of Von Neumann and Morgenstern utility functions for Pandora as she becomes more and more risk averse. In the limiting case, she cares only about not ending up with nothing at all. She will therefore be indifferent between $2 for certain and an even chance of $1 or $100. But the maximin criterion says that she should choose the former.

---

[8] Rawls's intuition in favor of the maximin criterion is perhaps better captured by John Milnor's (1954) theory of decisions under complete ignorance (section 9.1.1).

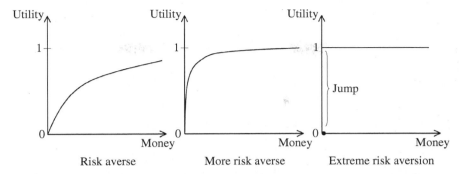

Figure 3.10. Extreme risk aversion. The graphs show Pandora's risk aversion increasing from left to right. In the extreme case, she is indifferent between all positive sums of money. She only cares about avoiding ending up with nothing at all.

The important point here isn't so much that the maximin criterion doesn't correspond to extreme risk aversion, but that it is incompatible with maximizing any kind of Von Neumann and Morgenstern utility function at all. It seems popular with some authors only because Von Neumann's theory of two-person zero-sum games recommends its use. But who would want to play a game against nature as though it were a zero-sum game in which nature's aim is to cause you as much damage as possible?

Risk neutrality. Pandora is risk neutral if she is always indifferent between having the dollar average of any set of money prizes for certain and participating in a lottery in which she has an equal chance of ending up with each prize. We then have equality in (3.4).

If Pandora is risk neutral, her Von Neumann and Morgenstern utility function can always be rescaled so that her utils can be identified with dollars. She therefore acts as though seeking to maximize her dollar profit. For example, insurance companies arguably behave as though they are risk neutral when insuring your house against fire. The reason is that they are also insuring very large numbers of other houses, and so the law of large numbers operates in their favor. If Pandora is risk averse, inequality (3.4) therefore explains why she can make a deal with an insurance company that profits both parties.

Suppose that Pandora's house is worth $$H$ and her other assets total $$W$. If the probability that the house burns down is r, then Pandora faces a lottery L in which she gets $W + H$ dollars with probability $1 - r$ and W with probability r. Inequality (3.4) tells us that she prefers having $W + H - rH$ dollars for certain to this lottery. So if the insurance company offers to insure her house for a premium not too much larger than

$P = rH$ dollars, Pandora will accept. However, $P = rH$ dollars is the break-even premium for a risk-neutral insurance company.

3.7.1 Taste for Gambling?

Pandora is risk loving if she always prefers participating in a lottery in which she has an equal chance of ending up with any one of a set of money prizes to having the average of the prizes for certain. The inequality in (3.4) is then reversed.

It is often taken for granted that gambling can be explained as rational behavior on the part of a risk-loving agent. Friedman and Savage (1948), for example, offer this explanation when trying to reconcile the behavior of people who both gamble and buy insurance.

The mistake is easily made, because to speak of "attitudes to risk" is a positive invitation to regard the shape of Pandora's Von Neumann and Morgenstern utility function u as embodying the thrill that she derives from the act of gambling. But if we fall into this error, we have no answer to the critics who ask why Von Neumann and Morgenstern utility functions should be thought to have any relevance to how Pandora chooses in riskless situations.

However, Von Neumann and Morgenstern's fourth postulate takes for granted that Pandora is entirely *neutral* about the actual act of gambling. She doesn't bet because she enjoys betting—she bets only when she judges that the odds are in her favor. If she liked or disliked the act of gambling itself, we would have no reason to assume that she is indifferent between a compound lottery and a simple lottery in which the prizes are available with the same probabilities. After all, it wouldn't be much fun to walk into a casino and bet all your money on one turn of the wheel at roulette. Gamblers choose instead to lose their money in dribs and drabs in return for the thrill of winning now and again.

To be rational in the sense of Von Neumann and Morgenstern, Pandora needs to be as unemotional about gambling as a Presbyterian minister of the old school when he insures his church against fire. Quentin may bet at the racetrack because he enjoys the excitement of the race. Rupert may refuse even to buy insurance because he believes any kind of gambling whatever is evil. Neither satisfy the Von Neumann and Morgenstern postulates, because they each like or dislike gambling for its own sake.

It doesn't follow that gambling for fun is necessarily irrational; only that the lotteries which give Pandora a thrill need to be excluded from those to which the Von Neumann and Morgenstern theory is applied. However, my guess is that most betting behavior is in fact irrational. Indeed, part of the reason that people enjoy gambling may be that they

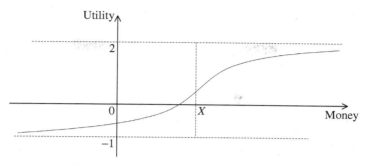

Figure 3.11. Bounded utility. This utility function is bounded above by 2 and below by −1. It is convex to the left of the point of inflexion X, and concave to the right.

like the feeling of behaving irrationally now and again. I recall, for example, suggesting to a regular loser at a weekly poker game that he keep a record of his winnings and losses. His response was that he used to do so but had given up because it proved to be unlucky.

3.8 Unbounded Utility?

The utility function drawn in figure 3.11 is bounded above by 2 and below by −1. The particular function drawn has a unique point of inflexion at X, and hence is concave to the right of X and convex to the left. It shares this property with the Friedman–Savage utility function of figure 3.8, but we have just seen that we aren't entitled to use this fact to explain why people gamble. On the other hand, to the small extent that the choice behavior of laboratory subjects can be explained in terms of maximizing expected utility functions, the functions that fit the data do seem to have such a unique point of inflexion (Kahneman and Tversky 1979).

My version of the Von Neumann and Morgenstern theory necessarily generates bounded utility functions because each lottery considered must lie between a best outcome W and a worst outcome $\mathcal{L}$. So what of the various examples involving money that we have been using? All of these involve unbounded utility functions.

Working with unbounded utility functions is unproblematic, provided we only do things that are sanctioned by Von Neumann and Morgenstern's postulates. What this means in practice is that we don't need to worry that a Von Neumann and Morgenstern utility function is unbounded if we only plan to consider lotteries whose expected utility is bounded. We can even allow lotteries with an infinite number of prizes if this constraint is observed.

What if we were to try to create a more glamorous theory in which we allow lotteries whose expected utility is infinite? Not only would we then be working outside the Von Neumann and Morgenstern theory, but we would have to face a whole morass of paradoxes.

Paradox of the infinite. Daniel Bernoulli's idea that people will maximize the expected value of a concave utility function doesn't make the St Petersburg paradox go away. Given any utility function u for money that is unbounded above, we can find a lottery with an infinite number of prizes whose expected utility is infinite.

To see why, simply replace the nth prize of $\$2^n$ in the St Petersburg lottery of figure 3.2 by a dollar prize $\mathcal{P}_n$ that is so large that $u(\mathcal{P}_n) \geqslant 2^n$ ($n = 1, 2, \dots$). The expected utility of the resulting lottery is then

$$u(\mathcal{P}_1)/2 + u(\mathcal{P}_2)/2^2 + u(\mathcal{P}_3)/2^3 + \cdots \geqslant 1 + 1 + 1 + \cdots.$$

Swapping envelopes. The devil offers Pandora a choice between two identical sealed envelopes, one of which is known to contain twice as much money as the other. Pandora chooses one of the envelopes and finds that it contains $\$2n$. So the other envelope contains either $\$n$ or $\$4n$. Pandora computes its expected dollar value to be

$$\tfrac{1}{2} \times n + \tfrac{1}{2} \times 4n = 5n/2 > 2n.$$

If she is risk neutral, Pandora will therefore always want to swap whatever envelope she chose for the other envelope.

The swapping envelopes paradox isn't so sharp as it may at first appear. There is nothing paradoxical about the fact that Pandora might prefer to swap whichever of two envelopes she opens first for the other. For example, Pandora might be offered two envelopes containing any adjacent pair of the prizes shown in figure 3.12(a). If she is offered the pair in the middle, then both envelopes will contain $\$1$, and she will want to swap whichever she opens for the other because it might possibly contain $\$2$ rather than $\$1$.

The paradox therefore lies in the fact that Pandora will wish to swap whichever envelope she chooses from *any* pair of envelopes she might be offered. Figure 3.12(b) illustrates that the devil can make Pandora want to swap whichever of two envelopes she is offered, except in the case when the pair of envelopes lies at one of the two extremes of the range of possibilities. It doesn't matter whether Pandora is risk neutral or not. Nor does it matter with what positive probability each pair of envelopes is offered. We only require that each prize has a neighbor to which Pandora assigns a large enough utility.

We can't make this argument work for every prize in the finite case, because the two extremal prizes will always be exceptions. But if we

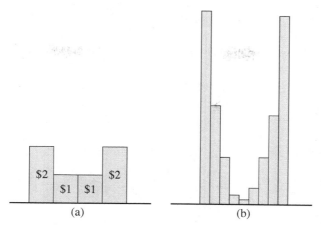

Figure 3.12. Swapping envelopes. Each of a pair of adjacent prizes is concealed in an envelope. Pandora chooses one of the two envelopes. Each prize not lying at an extreme has a neighbor whose utility is sufficiently large that Pandora wants to swap envelopes if she finds a nonextremal prize in the envelope she chose. If Pandora's utility function is unbounded, the extremes can be eliminated by allowing an infinite number of prizes.

allow an infinite number of prizes, we can eliminate the extremal cases altogether. If Pandora's Von Neumann and Morgenstern utility function is unbounded above, the devil can put sums of money in each of the infinite number of envelopes so as to guarantee that each prize has a neighbor with a sufficiently larger utility that Pandora always wants to swap whatever envelope she opens.

A subplot is sometimes allowed to confuse the issue. In the original story, Pandora always assigns probability $\frac{1}{2}$ to the other envelope containing either $\$n$ or $\$4n$. She is correct to do so only if the probability of each pair of envelopes that she might be offered is the same. But this can't be true in the infinite case, because an infinity of equal positive numbers must sum to infinity, and probabilities of nonoverlapping events can never sum to more than one. However, to focus on this error in the standard formulation of the paradox is to miss its essential point.

Pascal's wager. Infinite utilities are thought worthy of attention partly because of Pascal's wager. In this attempt to apply decision theory to theology, Pandora can choose to follow the straight and narrow path of rectitude, or she can indulge her passions. If there is an afterlife, the ultimate reward for living a good life will be infinitely more important than anything that might happen on this earth. Pascal's argument is therefore that Pandora ought to be good, even if she believes that the probability of an afterlife is very small.

It isn't interesting to challenge the prejudices built into Pascal's assumptions. Given his premises, the question is whether his argument is sound. The appearance of infinite magnitudes makes it clear that he can't appeal to the version of the Von Neumann and Morgenstern theory presented here in arguing that it is rational for Pandora to maximize her expected utility. If we don't allow infinite magnitudes, then all we learn is that it is rational for Pandora to be good if the probability of an afterlife isn't too small.

But is living a good life enough? God is commonly thought to demand belief in His existence as well as observance of His laws. It is said that adopting the good habits of believers is likely to result in your coming to believe as well. But to start living a good life for this reason is to fall foul of Aesop's principle.

3.9   Positive Applications?

Von Neumann and Morgenstern theory has been the subject of much debate in recent years. Are economists entitled to take for granted that the theory will predict the behavior of real people in risky situations? That is to say, is it safe to reinterpret the normative theory of Von Neumann and Morgenstern as a positive theory?

The chief exponents of the idea that it isn't safe are the psychologists Kahneman and Tversky (1979). Their laboratory experiments and those of their followers seem to me a knockdown refutation of the claim that the Von Neumann and Morgenstern theory is usually a good predictor of how ordinary people behave. For example, the indifference curves of someone who honors the Von Neumann and Morgenstern theory should all be parallel straight lines, as in figure 3.13. But experimental data shows most people reveal indifference curves that are neither straight nor parallel.

Various behavioral theories have been proposed as alternatives to the Von Neumann and Morgenstern theory (Machina 2004). The best known of these is Kahneman and Tversky's (1979) prospect theory. However, none of these rival theories are very successful in predicting the outcome of experiments whose results haven't been used already in calibrating the theory.

In fact, at the time of writing, the combatants in the debate seem largely to have retired from the scene after two papers appeared back-to-back in the journal *Econometrica* offering data which suggests that all extant positive theories are bad predictors—but that the least bad

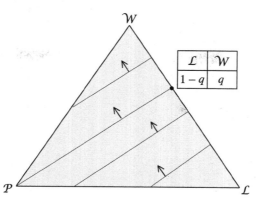

Figure 3.13. Rational indifference curves. Each point in the triangle represents a lottery with prizes L, P, and W. The lottery it represents lies at the center of gravity of weights located at each prize that are equal to the probability with which the prize is available. For example, the lottery Q in which Pandora gets L with probability $1 - q$ and W with probability q lies on the line segment joining L and W. If $q = \frac{2}{3}$, then Q is one third of the way down the line segment from W. Pandora is indifferent between P and the particular lottery Q shown in the figure. If we choose a Von Neumann and Morgenstern utility scale with $u(L) = 0$ and $u(W) = 1$, it follows that $u(P) = q$.

theory is the Von Neumann and Morgenstern theory of expected utility! (See Camerer and Harless (1994) and Hey and Orme (1994); and also Schmidt and Neugebauer (2007).)

My own views on when one might reasonably expect a normative theory of behavior to predict successfully in positive applications are given in section 1.10. Enthusiasts who somehow manage to convince themselves that the Von Neumann and Morgenstern theory always applies in every positive context only succeed in providing ammunition for skeptics looking for an excuse to junk the theory altogether. It is an unwelcome truth that the best we can say of positive applications of the theory is that it doesn't perform as badly overall as any of its behavioral rivals, but Aesop's principle tells us that it should be irrelevant whether a truth is welcome or not.

4
Utilitarianism

4.1 Revealed Preference in Social Choice

Until relatively recently, it was an article of faith among economists that one can't make meaningful comparisons of the utilities that different people may enjoy. This chapter is a good one to skip if you don't care about this question.

Social choice. The theory of social choice is about how groups of people make decisions collectively. This chapter is an aside on the implications of applying the theory of revealed preference to such a social context. Even in the case of a utilitarian government, we shall therefore be restricting our attention to notions of utility that make it fallacious to argue that policies are chosen because they maximize the sum of everyone's utility. On the contrary, we assign a greater utility to one policy rather than another because it is chosen when the other is available.

The theory of revealed preference is orthodox in individual choice, but not in social choice. However, no new concepts are required. If the choices made collectively by a society are stable and consistent, then the society behaves as though it were seeking to maximize something or other. The abstract something that it seems to be trying to maximize is usually called welfare rather than utility in a social context.

There are two reasons why it may be worth trying to figure out the welfare function of a society. The first is that we may hope to predict collective choices that the society may make in the future. The second is that bringing a society's welfare function to the attention of its citizens may result in their taking action to alter how their society works. For example, it may be that fitting a utilitarian welfare function to the available data is only possible if girls are treated as being worth only half as much as boys. Those of us who don't like sexism would then have a focus for complaint.

Difficulties with such an approach are easy to identify. What justifies the interpersonal comparisons of utility that are necessary for anything useful to emerge? Why should we imagine that even a rational society would have the consistent preferences we attribute to rational individuals?

4.1.1 Condorcet's Paradox

Thomas Hobbes' (1986) *Leviathan* treats a society like a single person written large. Karl Marx (Elster 1985) does the same for capital and labor. But a democratic society isn't a monolithic coalition that pursues its goals with the same single-minded determination as its individual citizens. The decisions it makes reflect a whole raft of compromises that are necessary to achieve a consensus among people with aspirations that will often be very different.

Condorcet's paradox of voting shows that a society which determines its communal preferences from the individual preferences of its citizens by honest voting over each pair of alternatives will sometimes reveal intransitive collective preferences. Suppose, for example, that the voting behavior of three citizens of a small society reveals that they each have transitive preferences over three alternatives:

$$a \prec_1 b \prec_1 c,$$
$$b \prec_2 c \prec_2 a,$$
$$c \prec_3 a \prec_3 b.$$

Then b wins a vote between a and b, c wins a vote between b and c, and a wins a vote between c and a. The collective preference so revealed is intransitive:

$$a \prec b \prec c \prec a.$$

In real life, voting is often strategic. For example, Alice may agree to vote with Bob on one issue if he will vote with her on another. Modern politics would founder altogether without such log-rolling agreements. But those of us who enjoy paradoxes need have no fear of being deprived of entertainment. Game theory teaches us that the conditions under which some version of the Condorcet paradox survives are very broad. However, the kind of welfarism we consider in this chapter assumes that stable societies can somehow manage to evade these difficulties.

4.1.2 Plato's Republic?

Ken Arrow (1963) generalized Condorcet's paradox from voting to a whole class of aggregation functions that map the individual preferences of the citizens to a communal preference. Arrow's paradox is sometimes said to entail that only dictatorial societies can be rational, but it represents no threat to utilitarianism because Arrow's assumptions exclude the kind of interpersonal comparison that lies at the heart of the utilitarian enterprise.

A dictatorial society in Arrow's sense is a society in which one citizen's personal preferences are always fully satisfied no matter what everyone else may prefer. The fact that utilitarian collective decisions are transitive makes it possible to think of utilitarian societies as being dictatorial as well, but the sense in which they are dictatorial differs from Arrow's, because the dictator isn't necessarily one of the citizens. He is an invented philosopher-king whose preferences are obtained by aggregating the individual preferences of the whole body of citizens. In orthodox welfare economics, the philosopher-king is taken to be a benign government whose internal dissensions have been abstracted away.

The imaginary philosopher-king of the welfarist approach is sometimes called an ideal observer or an impartial spectator in the philosophical literature, but the stronger kind of welfarism we are talking about requires that he play a more active role. Obedience must somehow be enforced on citizens who are reluctant to honor the compromises built into his welfare-maximizing choices.

Critics of utilitarianism enjoy inventing stories that cast doubt on this final assumption. For example, nine missionaries can escape the clutches of cannibals if they hand over a tenth missionary whose contribution to their mission is least important. The tenth missionary is unlikely to be enthusiastic about being sacrificed for the good of the community, but utilitarianism takes for granted that any protest he makes may be legitimately suppressed.

Authors who see utilitarianism as a system of personal morality tend to fudge this enforcement issue. For example, John Harsanyi (1977) tells us that citizens are morally committed to honor utilitarian decisions. John Rawls (1972) is implacably hostile to utilitarianism, but he too says that citizens have a natural duty to honor his version of egalitarianism. But what is a moral commitment or a natural duty? Why do they compel obedience? We get no answer, either from Harsanyi or from any of the other scholars who take this line.

My own view is that traditional utilitarianism only makes proper sense when it is offered, as in welfare economics, as a set of principles for a government to follow (Goodin 1995; Hardin 1988). The government then serves as a real enforcement agency that can effectively sanction "such actions as are prejudicial to the interests of others" (Mill 1962). However, just as it was important when discussing the rationality of individuals not to get hung up on the details of how the brain works, so we will lose our way if we allow ourselves to be distracted from considering the substantive rationality of collective decisions by such matters as the method of enforcement a society uses. It is enough for what will be said about utilitarianism that a society can somehow enforce the dictates of an imaginary philosopher-king.

Pandora's box? This last point perhaps needs some reinforcement. When we reverse the implicit causal relationships in traditional discussions by adopting a revealed-preference approach, we lose the capacity to discuss how a society works at the nuts-and-bolts level. What sustains the constitution? Who taxes whom? Who is entitled to what? How are deviants sanctioned? Such empirical questions are shut away in a black box while we direct our attention at the substantive rationality of the collective decisions we observe the society making.[1]

There is no suggestion that the social dynamics which cause a society to make one collective decision rather than another are somehow less important because we have closed the lid on them for the moment. As philosophers might say, to focus on consequences for a while is not to favor consequentialism over deontology. If we find a society—real or hypothetical—that acts as though it were maximizing a particular welfare function, the box will need to be opened to find out what social mechanisms are hidden inside that allow it to operate in this way. However, we are still entitled to feel pleased that temporarily closing our eyes to procedural issues creates a world so simple that we might reasonably hope to make sense of it.

4.2 Traditional Approaches to Utilitarianism

David Hume is sometimes said to have been the first utilitarian. He may perhaps have been the first author to model individuals as maximizers of utility, but utilitarianism is a doctrine that models a whole society as maximizing the sum of utilities of every citizen. This is why Bentham adopted the phrase "the greatest good for the greatest number" as the slogan for his pathbreaking utilitarian theory.

What counts as good? Bentham's attitude to this question was entirely empirical. The same goes for modern Benthamites who seek to harness neuroscience to the task of finding out what makes people happy (Layard 2005). Amartya Sen (1992, 1999) is no utilitarian, but his proposal to measure individual welfare in terms of human capabilities might also be

---

[1] The rationality of this book is sometimes called *substantive* rationality to emphasize that it is evaluated only in terms of consequences. One is then free to talk of *procedural* rationality when discussing the way that decisions get made.

said to be Benthamite in its determination to find measurable correlates of the factors that improve people's lives.

Bentham probably wouldn't have approved of a second branch that has grown from the tree he planted, because it takes for granted the existence of a metaphysical concept of the Good that we are morally bound to respect. John Harsanyi (1953, 1955) originated what I see as the leading twig on this metaphysical branch of utilitarianism. John Broome (1991), Peter Hammond (1988, 1992), and John Weymark (1991) are excellent references to the large literature that has since sprung into being.

John Harsanyi. I think that Harsanyi has a strong claim to be regarded as the true prophet of utilitarianism. Like many creative people, his early life was anything but peaceful. As a Jew in Hungary, he escaped disaster by the skin of his teeth not once, but twice. Having evaded the death camps of the Nazis, he crossed illegally into Austria with his wife to escape persecution by the Communists who followed. And once in the West, he had to build his career again from scratch, beginning with a factory job in Australia. It took a long time for economists to appreciate his originality, but he was finally awarded a Nobel prize shortly before his death. Moral philosophers are often unaware that he ever existed.

Harsanyi's failure to make much of an impression on moral philosophy is partly attributable to his use of a smattering of mathematics, but he also made things hard for himself by offering two different defenses of utilitarianism that I call his teleological and nonteleological theories (Binmore 1998, appendix 2).

4.2.1   Teleological Utilitarianism

What is the Good? G. E. Moore (1988) famously argued that the concept is beyond definition, but we all somehow know what it is anyway. Harsanyi (1977) flew in the face of this wisdom by offering axioms that supposedly characterize the Good. My treatment bowdlerizes his approach, but I think it captures the essentials.

The first assumption is that the collective decisions made by a rational society will satisfy the Von Neumann and Morgenstern postulates. A society will therefore behave as though it is maximizing the expected value of a Von Neumann and Morgenstern utility function. In this respect, the society behaves as though it were a rational individual whom I call the philosopher-king.

The second assumption is that the Von Neumann and Morgenstern utility function U of the philosopher-king depends *only* on the choices that the citizens would make if they were Arrovian dictators. If the N citizens of the society are no less rational than their philosopher-king, they will also behave as though maximizing the expected value of their own individual Von Neumann and Morgenstern utility functions u_n ($n = 1, 2, \ldots, N$). We then need to know precisely how the philosopher-king's utility function depends on the utility functions of the citizens.

To keep things simple, I suppose that $\mathcal{L}$ is a social state that everybody agrees is worse than anything that will ever need to be considered, and $\mathcal{W}$ is a social state that everyone agrees is better. We can use these extremal states to anchor everybody's utility scales so that $U(\mathcal{L}) = u_n(\mathcal{L}) = 0$ and $U(\mathcal{W}) = u_n(\mathcal{W}) = 1$. For each lottery L whose prizes are possible social states, we then assume that

$$\mathcal{E}U(L) = W(\mathcal{E}u_1(L), \mathcal{E}u_2(L), \ldots, \mathcal{E}u_N(L)),$$

so that the philosopher-king's expected utility for a lottery depends only on the citizens' expected utilities for the lottery.[2] The linearity of the expected utility functions then forces the function W to be linear (Binmore 1994, p. 281). But linear functions on finite-dimensional spaces necessarily have a representation of the form

$$W(x_1, x_2, \ldots, x_N) = w_1 x_1 + w_2 x_2 + \cdots + w_N x_N. \qquad (4.1)$$

If we assume that the philosopher-king will always improve the lot of any individual if nobody else suffers thereby, then the constants w_n ($n = 1, 2, \ldots, N$) will all be nonnegative.

Formula (4.1) is the definition of a utilitarian welfare function. The constants $w_1, w_2, \ldots, w_N$ that weight each citizen's individual utility determine the standard of interpersonal comparison built into the system. For example, if $w_3 = 2w_5$, then each util of citizen 3 is counted as being worth twice as much as each util of citizen 5. Figure 4.1 illustrates how the weighting influences the choices a utilitarian society will make from a given feasible set.

Harsanyi not a Benthamite? Amartya Sen (1976) argued that Harsanyi shouldn't be regarded as a utilitarian because he interprets utility in the sense of Von Neumann and Morgenstern. The criticism would be more justly directed at me for insisting on reinterpreting Harsanyi's work in terms of revealed-preference theory.

Harsanyi himself is innocent of the radicalism Sen attributes to him. Although he borrows the apparatus of modern utility theory, Harsanyi retains the Benthamite idea that utility causes our behavior. Where he genuinely differs from Sen is in having no truck with paternalism. Harsanyi's philosopher-king doesn't give people what is good for them

---

[2] I avoid much hassle here by assuming directly a result that other authors usually derive from what they see as more primitive axioms.

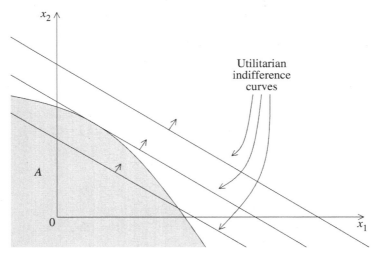

Figure 4.1. Utilitarian indifference curves. The diagram shows the indifference curves of the utilitarian welfare function $W(x_1, x_2) = x_1 + 2x_2$. The indifference curves are parallel straight lines for much the same reason that the same is true in figure 3.13. The point of tangency of an indifference curve to the boundary of the feasible set A is the outcome that maximizes welfare. If we were to increase player 2's weight from $w_2 = 2$ to $w_2 = 3$, her payoff would increase.

whether they like it or not. He tries to give them what they like, whether it is good for them or not.

4.3 Intensity of Preference

When is it true that $u(b) - u(a) > u(c) - u(d)$ implies that a person with the Von Neumann and Morgenstern utility function u would choose to swap b for a rather than c for d? Luce and Raiffa (1957) include the claim that Von Neumann and Morgenstern utility functions automatically admit such an interpretation in their list of common fallacies. But we have to be careful not to read too much into what Luce and Raiffa say.

It is true that nothing in the Von Neumann and Morgenstern postulates allows any deductions to be made about intensities of preference, because no assumptions about such intensities are built into the postulates. However, nothing prevents our making additional assumptions about intensities and following up the implications.

The intensity issue is especially important in a welfare context. The literature commonly assumes that most people have decreasing marginal utility: that each extra dollar assigned to Pandora is worth less to her than the previous dollar (section 3.3). But utilitarians don't want the same to be true of utils. They want each extra util that is assigned to Pandora to be worth the same to her as the previous util.

Von Neumann and Morgenstern's (1944, p. 18) take on the issue goes like this. Pandora reveals the preferences $a \prec b$ and $c \prec d$. We deem her to hold the first preference more intensely than the second if and only if she would always be willing to swap a lottery L in which the prizes a and d each occur with probability $\frac{1}{2}$ for a lottery M in which the prizes b and c each occur with probability $\frac{1}{2}$.

To see why Von Neumann and Morgenstern offer this definition of intensity, imagine that Pandora has a lottery ticket N that yields the prizes b and d with equal probabilities. Would she now rather exchange b for a in N, or d for c? Our intuition is that she would prefer the latter swap if and only if she thinks that b is a greater improvement on a than c is on d. But to say that she prefers the second of the two proposed exchanges to the first is to say that she prefers M to L.

In terms of Pandora's Von Neumann and Morgenstern utility function u, the fact that $L \prec M$ reduces to the proposition

$$\tfrac{1}{2}u(a) + \tfrac{1}{2}u(d) < \tfrac{1}{2}u(b) + \tfrac{1}{2}u(c).$$

So Pandora holds the preference $a \prec b$ more intensely than the preference $c \prec d$ if and only if $u(b) - u(a) > u(d) - u(c)$.

To evaluate the implications of Von Neumann and Morgenstern's definition of an intensity of preference, suppose that $u(b) = u(a) + 1$ and $u(d) = u(c) + 1$. Then Pandora gains 1 util in moving from a to b. She also gains 1 util in moving from c to d. Both these utils are equally valuable to her because she holds the preference $a \prec b$ with the same intensity as she holds the preference $c \prec d$.

4.4 Interpersonal Comparison of Utility

The main reason that Lionel Robbins (1938) and his contemporaries opposed the idea of cardinal utility is that they believed that to propose that utility could be cardinal was to claim that the utils of different people could be objectively compared (section 1.7). But who is to say that two people who respond similarly when put in similar positions are experiencing the same level of pleasure or pain in the privacy of their own minds? Since neuroscience hasn't yet reached a position where it might offer a reasonable answer, the question remains completely open.

Nor does it seem to help to pass from the Benthamite conception of utility as pleasure or pain to the theory of Von Neumann and Morgenstern. Attributing Von Neumann and Morgenstern utility functions to individuals doesn't automatically allow us to make interpersonal comparisons of utility. The claim that it does is perhaps the most important of the fallacies listed in Luce and Raiffa's (1957) excellent book. It is therefore unsurprising that economics graduate students are still often taught that interpersonal comparison of utility is *intrinsically* meaningless—especially since they can then be tricked into accepting the pernicious doctrine that outcomes are socially optimal if and only if they are Pareto-efficient.[3]

Von Neumann and Morgenstern (1944) clearly didn't think that interpersonal comparisons of utility are necessarily meaningless, since the portion of their *Theory of Games and Economic Behavior* devoted to cooperative game theory assumes not only that utils can be compared across individuals, but that utils can actually be transferred from one individual to another. The notion of transferable utility goes too far for me, since only physical objects can be physically passed from one person to another. However, as in the case of intensity of preferences, nothing prevents our adding extra assumptions to the Von Neumann and Morgenstern rationality postulates in order to make interpersonal comparisons meaningful.

In Harsanyi's teleological argument, the extra assumptions are those that characterize the choice behavior of a society, so that we can envisage it as being ruled by an all-powerful philosopher-king whose preferences are obtained by aggregating the preferences of his subjects. The standard of interpersonal comparison in the society is then determined by the personal inclinations of this imaginary philosopher-king from which we derive the weights in equation (4.1).

Where do the weights come from? According to revealed-preference theory, we estimate the weights by observing the collective choices we see a utilitarian society making. If its choices are stable and consistent, we can then use the estimated weights to predict future collective choices the society may make.

However, Harsanyi (1977) adopts a more utopian approach. He is dissatisfied with the idea that the standard of interpersonal comparison might be determined on paternalistic grounds that unjustly favor some individuals at the expense of others. He therefore requires that the philosopher-king be one of the citizens. The impartiality of a citizen elevated to the position of philosopher-king is guaranteed by having him make decisions as though behind a *veil of ignorance*. The veil of

---

[3] An outcome is Pareto-efficient if nobody can be made better off without making somebody else worse off. For example, it is Pareto-efficient to leave the poor to starve, because subsidizing them by taxing the rich makes rich people worse off.

ignorance conceals his identity so that he must make decisions on the assumption that it is equally likely that he might turn out be any one of the citizens on whose behalf he is acting.[4]

Whether or not such an approach to social justice seems reasonable, the method by which Harsanyi copes with the problem of interpersonal comparison within his theory is well worth some attention.

Anchoring points.   A temperature scale is determined by two anchoring points, which are usually taken to be the boiling point and the freezing point of water under certain conditions. Two anchoring points are also necessary to fix the zero and the unit on a Von Neumann and Morgenstern utility scale. In section 4.2, we took everybody's anchoring points to be $\mathcal{L}$ and $\mathcal{W}$.

It is sometimes argued that the mere choice of such common anchoring points solves the problem of interpersonal comparison. It is true that such a choice determines what counts as a util for each citizen. A society can therefore choose to trade off each util so defined against any other util at an equal rate. The weights in the utilitarian welfare function of (4.1) will then satisfy

$$w_1 = w_2 = \cdots = w_N.$$

The theory that results is called zero–one utilitarianism because the standard of interpersonal comparison is determined entirely by our decision to assign utility scales that set everybody's zero and one at $\mathcal{L}$ and $\mathcal{W}$. But what use is a standard chosen for no good reason? Who would want to use the zero–one standard if $\mathcal{W}$ and $\mathcal{L}$ meant that someone was to be picked at random to win or lose ten dollars when Adam is a billionaire and Eve is a beggar?

But the shortcomings of zero–one utilitarianism shouldn't be allowed to obscure the fact that comparing utils across individuals can be reduced to choosing anchoring points for each citizen's utility scale. To be more precise, suppose that citizen n reveals the preference $\mathcal{L}_n \prec_n \mathcal{W}_n$. Then $\mathcal{L}_n$ and $\mathcal{W}_n$ will serve as the zero and the unit points for a new utility scale for that citizen. The Von Neumann and Morgenstern utility function that corresponds to the new scale is given by

$$U_n(\mathcal{P}) = \frac{u_n(\mathcal{P}) - u_n(\mathcal{L}_n)}{u_n(\mathcal{W}_n) - u_n(\mathcal{L}_n)}, \tag{4.2}$$

where u_n is the citizen's old Von Neumann and Morgenstern utility function and U_n is the citizen's new Von Neumann and Morgenstern utility function.

---

[4] Moral philosophers will recognize a version of John Rawls's (1972) *original position*, which Harsanyi (1955) independently invented.

We can now create a standard of interpersonal comparison by decreeing that each citizen is to be assigned a weight to be used in comparing the utils on his or her new scale with those of other citizens. If we assign Alice a weight of 5 and Bob a weight of 4, then 4 utils on Alice's new scale are deemed to be worth the same as 5 utils on Bob's new scale.

Such a criterion depends on how the new anchoring points are chosen, but it doesn't restrict us to using the new utility scales. For example, reverting to Alice's old utility scale may require replacing her new Von Neumann and Morgenstern utility function U by $u = 2U + 3$. We must then accept that 8 utils on Alice's old scale are to be deemed to be worth the same as 5 utils on Bob's new scale.

Such mathematical reasoning tells us something about the structure of any standard of interpersonal comparison of utility, but it can't tell us anything substantive about whether some particular standard makes sense in a particular context. We need some empirical input for this purpose. For example, how do people in a particular society feel about being rich or poor? How do they regard the sick and the lame? In his nonteleological argument for utilitarianism, Harsanyi followed David Hume and Adam Smith in looking to our capacity to empathize with others for input on such questions.

4.4.1 Empathetic Preferences

If Quentin's welfare appears as an argument in Pandora's utility function, psychologists say that she *sympathizes* with him. For example, a mother commonly cares more for her baby's welfare than for her own. Lovers are sometimes no less unselfish. Even Milton Friedman apparently derived a warm glow from giving a small fraction of his income to relieve the distress of strangers in faraway places. Such sympathetic preferences, however attenuated, need to be distinguished from the empathetic preferences to be discussed next.[5]

Pandora *empathizes* with Quentin when she puts herself in his position to see things from his point of view. Autism in children often becomes evident because they commonly lack this talent. Only after observing how their handicap prevents their operating successfully in a social context does it become obvious to what a large degree the rest of us depend on our capacity for empathy.

Pandora expresses an empathetic preference if she says that she would rather be Quentin enjoying a quiet cup of coffee than Rupert living it up

---

[5] Economists traditionally follow the philosopher Patrick Suppes (1966) in referring to empathetic preferences as extended sympathy preferences.

with a bottle of champagne. Or she might say that she would like to swap places with Sally, whose eyes are an enviable shade of blue.

To hold an empathetic preference, you need to empathize with what others want, but you may not sympathize with them at all. For example, we seldom sympathize with those we envy, but Alice can't envy Bob without comparing their relative positions. However, for Alice to envy Bob, it isn't enough for her to imagine having Bob's possessions and her own preferences. Even if she is poor and he is rich, she won't envy him if he is suffering from incurable clinical depression. She literally wouldn't swap places with him for a million dollars. When she compares her lot with his, she needs to imagine how it would be to have both his possessions *and his preferences*. Her judgment on whether or not to envy Bob after empathizing with his full situation will be said to reveal an empathetic preference on her part.

Modeling empathetic preferences. In modeling the empathetic preferences that Pandora reveals, we retain as much of the apparatus from previous chapters as we can. In particular, the set of outcomes to be considered is lott(X), which is the set of all lotteries over a set X of prizes. We write $L \prec M$ when talking about the personal preferences revealed by Pandora's choice behavior. We write $(L, \ell) \prec (M, m)$ when talking about Pandora's empathetic preferences. Such a relation registers that Pandora has revealed that she would rather be citizen m facing lottery M than citizen ℓ facing lottery L.

Harsanyi appends two simple assumptions about Pandora's empathetic preferences to those of Von Neumann and Morgenstern.

Postulate 5. Empathetic preference relations satisfy all the Von Neumann and Morgenstern postulates.

Pandora's empathetic preference relation can therefore be represented by a Von Neumann and Morgenstern utility function v that assigns a utility of

$$v_n(\mathcal{P}) = v(\mathcal{P}, n)$$

to the event that she finds herself in the role of citizen n facing situation $\mathcal{P}$.

Harsanyi motivates the next postulate by asking how Pandora will decide to whom she should give a hard-to-get opera ticket that she can't use herself. One consideration that matters is whether Alice or Bob will enjoy the performance more. If Bob prefers Wagner to Mozart, then the final postulate insists that Pandora will choose Wagner over Mozart when making judgments on Bob's behalf, even though she herself may share my distaste for Wagner. No paternalism at all is therefore allowed to intrude into Harsanyi's system.

Postulate 6. Pandora is fully successful in empathizing with other citizens. When she puts herself in Quentin's shoes, she accepts that if she were him, she would make the same personal choices as he would.

The postulate is very strong. For example, it takes for granted that Pandora knows enough about the citizens in her society to be able to summarize their choice behavior using personal Von Neumann and Morgenstern utility functions. When attempts to agree on what is fair in real life go awry, it must often be because this assumption fails to apply.

In precise terms, postulate 6 says that if u_n is a Von Neumann and Morgenstern utility function that represents citizen n's personal preferences, and v is a Von Neumann and Morgenstern utility function that represents Pandora's empathetic preferences, then $\mathcal{E}v_n$ represents the same preference relation on $\text{lott}(X)$ as $\mathcal{E}u_n$.

It is convenient to rescale citizen n's Von Neumann and Morgenstern utility by replacing u_n by U_n, as defined by (4.2). We then have that $U_n(\mathcal{L}_n) = 0$ and $U_n(\mathcal{W}_n) = 1$. We can choose any utility scale to represent Pandora's empathetic preferences, provided we don't later forget which anchoring points we chose to use.

Two Von Neumann and Morgenstern utility functions that represent the same preferences over $\text{lott}(X)$ are strictly increasing affine transformations of each other (section 3.6). So we can find constants $a_n > 0$ and b_n such that

$$v(\mathcal{P}, n) = a_n U_n(\mathcal{P}) + b_n \tag{4.3}$$

for any prize $\mathcal{P}$ in X. We can work out a_n and b_n by substituting $\mathcal{L}_n$ and $\mathcal{W}_n$ for $\mathcal{P}$ in (4.3) and then solving the resulting simultaneous equations. We find that

$$a_n = v(\mathcal{W}_n, n) - v(\mathcal{L}_n, n),$$
$$b_n = v(\mathcal{L}_n, n).$$

The answers we get depend on Pandora's empathetic utilities at citizen n's anchoring points, because it is her empathetic preferences that we are using to determine a standard of interpersonal comparison of utility.[6]

We now propel Pandora into the role of a philosopher-king behind Harsanyi's veil of ignorance. Which lottery L will she choose from whatever feasible set she faces? If she believes that she will end up occupying the role of citizen n with probability p_n, she will choose as though

---

[6] It makes no effective difference if we replace v by $Av + B$, where $A > 0$ and B are constants.

maximizing her expected utility:

$$p_1 \mathcal{E}v(L, 1) + \cdots + p_N \mathcal{E}v(L, N) = p_1 a_1 \mathcal{E}U_1(L) + \cdots + p_N a_N \mathcal{E}U_N(L),$$

where the constant term $p_1 b_1 + p_1 b_1 + \cdots + p_1 b_1$ has been suppressed because its value makes no difference to Pandora's choice. If we write $x_n = \mathcal{E}U_n(L)$, we find that Pandora then acts as though maximizing the utilitarian welfare function of (4.1) in which the weights are given by

$$w_n = p_n a_n.$$

In summary: when Pandora makes impartial choices on behalf of others, she can be seen as revealing a set of empathetic preferences. With Harsanyi's assumptions, these empathetic preferences can be summarized by stating the rates at which she trades off one citizen's utils against another's.

The Harsanyi doctrine. The preceding discussion avoids any appeal to what has become known as the Harsanyi doctrine, which says that rational people put into identical situations will necessarily make the same decisions.

In Harsanyi's (1977) full nonteleological argument, all citizens are envisaged as bargaining over the collective choices of their society behind a veil of ignorance that conceals their identities during the negotiation. This is precisely the situation envisaged in Rawls's (1972) original position, which is the key concept of his famous *Theory of Justice.* Harsanyi collapses the bargaining problem into a one-person decision problem with a metaphysical trick. Behind his veil of ignorance, the citizens forget even the empathetic preferences they have in real life. They are therefore said to construct them anew. The Harsanyi doctrine is then invoked to ensure that they all end up with the same empathetic preferences. So their collective decisions can be delegated to just one of their number, whom I called Pandora in the preceding discussion of empathetic preferences.

Harsanyi's argument violates Aesop's principle at a primitive level. If you could choose your preferences, they wouldn't be preferences but actions! However, Rawls is far more iconoclastic in his determination to avoid the utilitarian conclusion to which Harsanyi is led. He abandons orthodox decision theory altogether, arguing instead for using the maximin criterion in the original position (section 3.7).

My own contribution to this debate is to draw attention to the relevance of the enforcement issue briefly mentioned in section 4.1. If one gives up the idea that something called "natural duty" enforces hypothetical deals made in the original position, one can use game theory to defend a version of Rawls's egalitarian theory (Binmore 1994, 1998, 2005). My position on the debate between Harsanyi and Rawls about rational bargaining in the original position is therefore that Harsanyi was the better analyst but Rawls had the better intuition.

5

Classical Probability

5.1 Origins

Probability doesn't come naturally to the human species. The ancients never came up with the idea at all, although they enjoyed gambling games just as much as we do. It was only in the seventeenth century that probability saw the light of day. The Chevalier de Méré will always be remembered for proposing a problem about gambling odds that succeeded in engaging the attention of two great mathematicians of the day, Pierre de Fermat and Blaise Pascal. Some of the letters they exchanged in 1654 still survive.[1] It is fascinating to learn how difficult they found ideas that we teach to school children as though they were entirely unproblematic.

Ian Hacking's (1975) *Emergence of Probability* documents how the pioneering work of Fermat and Pascal was pursued by a galaxy of famous mathematicians, including Jacob Bernoulli, Huygens, Laplace and de Moivre. It is generally thought that the classical theory of probability was finally brought to a state of perfection in 1933 when Kolmogorov (1950) formulated the system of axioms that bears his name.

We took Kolmogorov's formulation of probability theory for granted when presenting Von Neumann and Morgenstern's theory of expected utility (section 3.4). However, in seeking a theory of decision that applies more widely, I need to explore the implicit assumptions built into Kolmogorov's axioms. In taking on this task, the current chapter says nothing that mathematicians or statisticians will find unorthodox, but its focus is on interpretive questions that aren't normally discussed very much, rather than on the mathematics of the theory.

5.2 Measurable Sets

ups!

The prosaic attitude of statisticians to probability is reflected in the language they use. In this book, the set B in a decision problem

---

[1] The spoof letters between Pascal and Fermat written by Alfréd Rényi (1977) are more entertaining than the real ones. Rényi's potted history of the early days of probability is also instructive.

$D : A \times B \rightarrow C$ is usually said to be the set of states of the world. Philosophers sometimes call B the universe of discourse. Statisticians call B a sample space.

The subsets of a sample space B are identified with possible events. For example, the sample space when a roulette wheel is spun is usually taken to be the set $\{0, 1, 2, 3, \ldots, 36\}$ of possible numbers that may come up.[2] The event that the number is odd is the subset $\{1, 3, 5, \ldots, 35\}$.

It hasn't mattered very much so far whether the set B is finite or infinite, but from now on it will be important not to exclude the possibility that it is infinite. Allowing B to be infinite creates various problems that some authors seek to exclude by insisting that only finite sets exist in the real world. Perhaps they are right, but our models aren't the real world—they are always a draconian simplification of whatever the real world may be. Allowing infinite sets into our models is often part of the process of simplification, as when Isaac Newton modeled space and time as continuous variables to which his newly minted calculus could be applied.

In decision theory, there are even more pressing reasons for allowing an infinite number of states of the world. For example, the world we are trying to model in game theory sometimes includes the minds of the other players. When thinking about how to play, they may do some calculations. Do we want to restrict them to using only numbers smaller than some upper bound? When advising Alice, do we want to assume that her model of the world is sufficient to encompass all the models that Bob might be using, but not that Bob's model is sufficient to encompass all the models that Alice might be using?

The last point takes us into even deeper waters. Do we really want to proceed as though we are capable of making a finite list of all relevant possibilities in any situation that might conceivably arise? I think this is demonstrably impossible even if we allow ourselves an infinite number of states (section 8.4). As Hamlet says to Horatio: "There are more things in heaven and earth, Horatio, than are dreamt of in your philosophy." In brief, we need to avoid fooling ourselves into thinking that we always know how to scale down the universe into a manageable package.

The latter problem is captured in classical probability theory by saying that some events are measurable and others are not. The measurable events are those that we can tie down sufficiently to make it meaningful

---

[2] American roulette wheels have a slot labeled 00, so they are more unfair than European wheels, since you are still only paid as though the probability of any particular slot coming up is $1/36$.

to attach a probability to them. To speak of the probability of a nonmeasurable set is to call upon the theory to deliver something for which it is unequipped.

In later chapters, I shall be arguing that one can't follow the traditional practice of ignoring the possible existence of nonmeasurable sets when seeking to make rational decisions in a large world. However, it is difficult to get a feeling for the extent to which nonmeasurable sets can differ from their measurable cousins when they are only discussed in the abstract. It is therefore worth reviewing some of the examples that arise in Euclidean geometry.

5.2.1 Lebesgue Measure

Henri Lebesgue created the subject of measure theory when he showed how the concept of length can be extended to the class of Lebesgue measurable sets on the real line. In offering a potted explanation of his idea, I shall talk about the length of arcs on a given circle rather than the length of intervals on a line, because it is then possible to make the measure that results into a probability measure by normalizing so that the arc length of the whole circle is one. If you like, you can think of the circle as a generalized roulette wheel.

Lebesgue's basic idea is simple. He saw that one can always use any collection C of measurable sets on a circle—sets to which one has already found a way of attaching an arc length—to find more measurable sets.

Suppose that S is contained in a set T in the collection C. If S is measurable, its measure $m(S)$ can't be more than $m(T)$. In fact,

$$m(S) \leqslant \overline{m}(S), \qquad (5.1)$$

where $\overline{m}(S)$ is the largest real number no larger than all $m(T)$ for which $S \subseteq T$.

Similarly, if S contains a set T in the collection C, then $m(S)$ can't be less than $m(T)$. In fact,

$$m(S) \geqslant \underline{m}(S), \qquad (5.2)$$

where $\underline{m}(S)$ is the smallest real number no smaller than all $m(T)$ for which $T \subseteq S$.

For some sets S, it will be true that

$$\overline{m}(S) = \underline{m}(S). \qquad (5.3)$$

It then makes sense to say that S is measurable, and that its Lebesgue measure $m(S)$ is the common value of $\overline{m}(S)$ and $\underline{m}(S)$.

In the case of Lebesgue measure on a circle, it is enough to begin with a collection C whose constituent sets are all unions of nonoverlapping open arcs. The measure of such a set is taken to be the sum of the lengths of all the arcs in this countable collection.[3] The sets that satisfy (5.3) with this choice of C exhaust the class $\mathcal{L}$ of all Lebesgue measurable sets on the circle. Replacing C by $\mathcal{L}$ in an attempt to extend the class of measurable sets just brings us back to $\mathcal{L}$ again.

What do Lebesgue measurable sets look like? Roughly speaking, the answer is that any set that can be specified with a formula that uses the standard language of mathematics is Lebesgue measurable.[4] Nonmeasurable sets are therefore sets that our available language is inadequate to describe. But do such sets really make sense in concrete situations?

Vitali's nonmeasurable set.   Vitali's proof of the existence of a set on the circle that isn't Lebesgue measurable goes like this.

We first say that two points on the circle are to be included in the same class if the acute angle they subtend at the center is a rational number.[5] There are many such equivalence classes, but they never have a point in common. The next step is to make a set P by choosing precisely one point from each class. We suppose that P has measure m and seek a contradiction.

Let P_a be the set obtained by rotating P around the center of the circle through an acute angle a. The construction ensures that the collection of all P_r for which r is a rational number *partitions* the circle. This means that no two sets in the collection overlap, and that its union is the whole circle. It follows that the measure of the whole circle is equal to the sum of the measures of each member of the partition.

But all the sets in the partition are rotations of P, and so have the same measure m as P. The measure of the whole circle must therefore either be zero if $m = 0$, or infinity if $m > 0$. Since the measure of the whole circle is one, we have a contradiction, and so P is nonmeasurable.

The Horatio principle.   The nonmeasurable set P is obtained by picking a point from each of the equivalence classes with which the proof begins, but no hint is offered on precisely how each point is to be chosen. In fact, to justify this step in the proof, an appeal needs to be made to the axiom of choice, which set theorists invented for precisely this kind of purpose. But is the axiom of choice true?

---

[3] There can only be a countable number of arcs in the collection, because each arc must contain a distinct point that makes a rational angle with a fixed radius. But there are only a countable number of rational numbers (section 1.7).

[4] The full answer is that any Lebesgue measurable set is such a Borel set after removing a set of measure zero.

[5] An acute angle lies between 0° and 180° inclusive.

I think it fruitless to ask such metaphysical questions. The real issue is whether a model in which we assume the axiom of choice better fits the issues we are seeking to address than a model in which the axiom of choice is denied. We are free to go either way, because it is known that the axiom of choice is independent of the other axioms of set theory. Moreover, if we deny the axiom of choice, Robert Solovay (1970) has shown that we can then consistently assume that all sets on the circle are Lebesgue measurable.[6]

My own view is that to deny the axiom of choice in a large-world setting is to abandon all pretence at taking the issues seriously. It can be interpreted as saying that nature has ways of making choices that we can't describe using whatever language is available to us. In a sufficiently complex world, the implication is that some version of what I shall call the Horatio principle must apply:

Some events in a large world are necessarily nonmeasurable.

The Banach–Tarski paradox. The properties of nonmeasurable sets become even more paradoxical when we move from circles to spheres (Wagon 1985).

Felix Hausdorff showed that we can partition the sphere into three sets, each of which can be rotated onto any one of the others. There is nothing remarkable in this, but each of his sets can also be rotated onto the union of the other two! If one of Hausdorff's sets were Lebesgue measurable, its measure would therefore need to be equal to twice itself.

The Banach–Tarski paradox is even more bizarre. It says that a sphere can be partitioned into a finite number of nonoverlapping subsets that can be reassembled to make two distinct spheres, each of which is identical to the original sphere. Perhaps this is how God created Eve from Adam's rib!

5.3 Kolmogorov's Axioms

Kolmogorov's formulation of classical probability theory is entirely abstract. Measurable events are simply sets with certain properties. A probability measure is simply a certain kind of function. Kolmogorov doesn't insist on any particular interpretation of his theory, whether metaphysical or otherwise. His attitude is that of a pure mathematician. If we can find a model to which his theory applies, then we are entitled to apply all of his theorems in our model.

---

[6] If we are also willing to postulate the existence of "inaccessible cardinals."

Abstract measurable sets.   Kolmogorov first requires that one can never create a nonmeasurable set by putting his abstract measurable events together in various ways using elementary logic. He therefore makes the following assumptions.

1. *The whole sample space is measurable.*

2. *If an event is measurable, then so is its complement.*

3. *If each of a finite collection of events is measurable, then so is their union.*

I have listed these formal requirements only because Kolmogorov's theory requires that the third item be extended to the case of infinite but countable unions:

3*. *If each of a* countable *collection of events is measurable, then so is their union.*

A set is countable if it can be counted. To count a set is to arrange it in a sequence, so that we can assign 1 to its first term, 2 to its second term, 3 to its third term, and so on. The infinite set of all natural numbers can obviously be counted. We used the fact that the set of all rational numbers is countable in section 1.7. The set $\mathbb{R}$ of all real numbers is uncountable. No matter how we try to arrange the real numbers in a sequence, there will always be real numbers that are left out.

A mundane nonmeasurable set.   After seeing Vitali's construction of a set on the circle that isn't Lebesgue measurable, it is natural to think that a nonmeasurable set must be a very strange object indeed, but Kolmogorov's definition allows the most innocent of sets to be nonmeasurable.

Suppose, for example, that Pandora tosses a coin. The coin may have been tossed many times before and perhaps will be tossed many times again, but we don't know anything about the future or the past. If we took our sample space B to be all finite sequences of *heads* and *tails*, with Pandora's current toss located at some privileged spot in the sequence, then we might take account of our ignorance by only allowing the measurable sets to be $\varnothing$, H, T, and B, where H is the event that Pandora's current toss is *heads* and T is the event that her current toss is *tails*.

The event that some previous toss of the coin was *heads* is then nonmeasurable in this setup. Of course, other models could be constructed in which the same event is measurable. As with so much else, such considerations are all relative to the model one chooses to adopt.

5.3.1 Abstract Probability

Kolmogorov defines a probability space to be a collection $\mathcal{F}$ of measurable events from a sample space B together with a probability measure

$$\text{prob} : \mathcal{F} \to \mathbb{R}.$$

It will be important not to forget that the theory only defines probabilities on measurable sets.[7] The requirements for a probability measure are:

1. *For any E, $\text{prob}(E) \geqslant 0$ with equality for the impossible event $E = \varnothing$.*

2. *For any E, $\text{prob}(E) \leqslant 1$ with equality for the certain event $E = B$.*

3. *If no two of the countable collection of events $E_1, E_2, \ldots$ have a state in common, then the probability of their union is*

$$\text{prob}(E_1) + \text{prob}(E_2) + \text{prob}(E_3) + \cdots.$$

The union corresponds to the event that at least one of the events $E_1, E_2, \ldots$ occurs. Item 3 therefore says that we should add the probabilities of exclusive events to get the probability that one of them will occur.

Countable additivity? The third requirement for a probability measure is called countable additivity. It is the focus of much debate among scholars who either hope to generalize classical probability theory in a meaningful way, or else feel that they know the one-and-only true interpretation of probability. The former seek to extend the class of probability functions by exploring the implications of replacing countable additivity by finite additivity. The latter class of scholars has a distinguished representative in the person of Bruno de Finetti (1974b, p. 343), who tells us in capital letters that he REJECTS countable additivity.

Although de Finetti was a man of genius, we need not follow him in rejecting countable additivity for what seem to me metaphysical reasons. But nor do we need to hold fast to countable additivity if it stands in our way. Different applications sometimes require different models. Where we can't have countable additivity, we must make do with finite additivity.[8]

---

[7] Formally, a probability space is a triple $(B, \mathcal{F}, p)$, where B is a sample space, $\mathcal{F}$ is a collection of measurable sets in B, and p is a probability measure defined on $\mathcal{F}$. Mathematicians express Kolmogorov's measurability requirements by saying that $\mathcal{F}$ must be a σ-algebra.

[8] Nicholas Bingham (2005) provides an excellent historical survey of the views of mathematicians on finite additivity.

There is an unfortunate problem about nomenclature when probabilities are allowed to be only finitely additive. A common solution is to talk about finitely additive measures, disregarding the fact that measures are necessarily countably additive by definition. The alternative is to invent a new word for a finitely additive function μ for which $\mu(\varnothing) = 0$. I somewhat reluctantly follow Rao and Rao (1983) and Marinacci and Montrucchio (2004) in calling such a function a *charge*, but without any intention of using more than a smattering of their mathematical developments of the subject.

However, abandoning countable additivity is a major sacrifice. Useful infinite models can usually be regarded as limiting cases of the complex finite models that they are intended to approximate. In such cases, we don't want "discontinuities at infinity" as a mathematician might say. The object in an infinite model that supposedly approximates an object in a complex finite model must indeed be an approximation to that object. In the case of probability theory, such a continuity requirement can't be satisfied unless we insist that the probability of the limit of a sequence of measurable sets is equal to the limit of their individual probabilities.[9] But countable additivity reduces to precisely this requirement.

This is only one reason that enthusiasts for finite additivity who think that junking countable additivity will necessarily make their lives easier are mistaken. We certainly can't eliminate the problem of nonmeasurable sets in this way. It is true that we can find a finitely additive extension of Lebesgue measure to all sets on a circle, but Hausdorff's paradox of the sphere shows that the problem returns to haunt us as soon as we move up to three dimensions.

5.4 Probability on the Natural Numbers

When people say that a point is equally likely to be anywhere on a circle, they usually mean that its location is uniformly distributed on the circle. This means that the probability of finding the point in a Lebesgue measurable set E is proportional to the Lebesgue measure of E. If E isn't measurable, we are able to say nothing at all.

It is a perennial problem that classical probability theory doesn't allow a uniform distribution on the real line. There is equally no probability measure p defined on the set $\mathbb{N}$ of natural numbers with $p(1) = p(2) = p(3) = \cdots$. If there were, then

$$p(\mathbb{N}) = p(1) + p(2) + p(3) + \cdots$$

---

[9] One only need consider sequences of sets that get larger and larger. Their limit can then be identified with their union.

would have to be either 0 or $+\infty$, but $p(\mathbb{N}) = 1$. It is sometimes suggested that we should abandon the requirement that $\mathrm{prob}(\mathbb{N}) = 1$ in order to escape this problem, but I think we would then throw the baby out with the bathwater.

De Finetti's (1974a, 1974b) solution is bound up with his rejection of countable additivity. With finite additivity one can have $0 = p(1) = p(2) = p(3) = \cdots$ without having to face the contradiction $1 = 0 + 0 + 0 + \cdots$ (Kadane and O'Hagan 1995).

I don't share de Finetti's enthusiasm for rejecting countable additivity, but one can't but agree that we have to do without it when situations arise in which each natural number needs to be treated as equally likely. However, de Finetti wouldn't have approved at all of my applying the idea to objective probabilities in the next chapter.

5.5 Conditional Probability

Kolmogorov introduces conditional probability as a definition:

$$\mathrm{prob}(E\,|\,F) = \frac{\mathrm{prob}(E \text{ and } F)}{\mathrm{prob}(F)}. \tag{5.4}$$

Thomas Bayes' famous rule follows immediately:

$$\mathrm{prob}(E\,|\,F)\,\mathrm{prob}(F) = \mathrm{prob}(F\,|\,E)\,\mathrm{prob}(E). \tag{5.5}$$

Base-rate fallacy. The following problem illustrates the practical use of Bayes' rule. A disease afflicts one person in a hundred. A test for the disease gets the answer right 99% of the time. If you test positive, what is the probability that you have the disease? To calculate the answer from Bayes' rule, let I be the event that you are ill and P the event that you test positive. We then have that

$$\mathrm{prob}(I\,|\,P) = \frac{\mathrm{prob}(P\,|\,I)\,\mathrm{prob}(I)}{\mathrm{prob}(P)} = \frac{0.99 \times 0.01}{0.99 \times 0.01 + 0.01 \times 0.99} = \frac{1}{2},$$

where the formula $\mathrm{prob}(P) = \mathrm{prob}(P\,|\,I)\,\mathrm{prob}(I) + \mathrm{prob}(P\,|\,{\sim}I)\,\mathrm{prob}({\sim}I)$ used to work out the denominator also follows from the definition of a conditional probability.

Most people are surprised that they only have a 50% chance of being ill after testing positive, because we all have a tendency to ignore the base rate at which people in the population at large catch the disease. The modern consensus seems to be that Kahneman and Tversky (1973) overestimated the effect to which ordinary people fall prey to the base-rate fallacy, but it is true nevertheless that our natural ineptitude with probabilities gets much worse when we have to think about conditional probabilities (section 6.3.3).

Hempel's paradox.   This paradox illustrates the kind of confusion that can arise when conditional probabilities appear in an abstract argument.

Everybody agrees that observing a black raven adds support to the claim that all ravens are black.[10] Hempel's paradox says that observing a pink flamingo should also support the claim that all ravens are black. But what possible light can sighting a pink flamingo cast on the color of ravens?

The answer we are offered is that pink flamingos are neither ravens nor black, and "*P* implies *Q*" is equivalent to "(not *Q*) implies (not *P*)." The logic is impeccable, but what relevance does it have to the issue? In such situations, it always helps to write down a specific model that is as simple as possible.

The model illustrated in figure 5.1 assumes that there are only three birds in the world, which have been represented by stars. Two of the birds are ravens and the other isn't. The squares are therefore our states of the world. They show all the possible ways the three birds might be colored if we only distinguish between black and not black. Each square is labeled with its prior probability—the probability we assign to the state prior to making an observation.

To say that all ravens are black is to assert that the event E has occurred. Before any observation, $\text{prob}(E) = h_{11} + h_{21}$. To observe a black raven is to learn that the event F has occurred. What is $\text{prob}(E \mid F)$? Applying Kolmogorov's definition (5.4), we find that

$$\text{prob}(E \mid F) = \frac{\text{prob}(E \text{ and } F)}{\text{prob}(F)} = \frac{h_{11} + h_{21}}{h_{11} + h_{12} + h_{21} + h_{22}}.$$

It follows that observing a black raven never decreases the probability that all ravens are black. It increases the probability unless $h_{13} = h_{23} = 0$ (which is the condition that no ravens are black).

To observe a pink flamingo is to learn that the event G has occurred. Then,

$$\text{prob}(E \mid G) = \frac{\text{prob}(E \text{ and } G)}{\text{prob}(G)} = \frac{h_{11}}{h_{11} + h_{12} + h_{13}}.$$

Observing a pink flamingo may therefore increase or decrease the probability that all ravens are black depending on the distribution of the prior probabilities. Nor do matters change in this respect if we examine a more realistic model with a large number of birds.

---

[10] If $E \subseteq F$ and $0 < \text{prob}(F) < 1$, then it follows from the definition of a conditional probability that $\text{prob}(E \mid F) > \text{prob}(E)$.

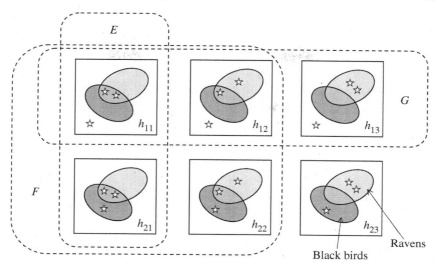

Figure 5.1. Hempel's paradox. The stars represent three birds. Two of the birds are ravens and the other isn't. Each square represents a possible state of the world labeled with its probability before any observation is made. The event E corresponds to the claim that all ravens are black. After observing a black raven, we learn that the event F has occurred. After observing a pink flamingo, we learn that the event G has occurred. What are $\text{prob}(E\,|\,F)$ and $\text{prob}(E\,|\,G)$?

5.5.1 Justifying the Formula?

The confusion commonly engendered by Hempel's paradox and our vulnerability to the base-rate fallacy suggests that we ought to think hard about Kolmogorov's definition of a conditional probability when we seek to make use of his abstract mathematical theory in a particular context.

It is easy to justify Kolmogorov's definition in the objective case when probability is to be interpreted as a naturally occurring frequency, like the rate at which uranium atoms decay. It is harder to justify the definition in the subjective case when probability is interpreted in terms of the odds at which someone is willing to bet on various events. It is still harder to find reasons why one should use Kolmogorov's definition in the logical case when probability is to be interpreted as the rational degree of belief given the available evidence.

These interpretive issues are discussed at length in later chapters, but it is important at this stage to emphasize that the Kolmogorov theory itself offers no source of authority to those who advocate one interpretation of probability as opposed to another. The predictive success of the theory in physics certainly provides immense empirical support for the frequency interpretation where this applies, but one can't parlay this into support for the aptness of Kolmogorov's definition of a conditional probability when probability isn't interpreted as a frequency.

Since Bayes' rule is deduced from Kolmogorov's definition of a conditional probability by a trivial algebraic manipulation, the same goes for Bayesian updating too. It isn't written into the fabric of the universe that Bayes' rule is the one-and-only correct way to update our beliefs. The naive kind of Bayesianism in which this assumption isn't even questioned lacks any kind of foundation at all. A definition is just a definition and nothing more.

5.5.2 Independence

As in the case of conditional probability, Kolmogorov's approach to independence is simply to offer a definition. Two events E and F are defined to be independent if the probability that they will both occur is the product of their probabilities. It then follows that

$$\text{prob}(E \,|\, F) = \text{prob}(E),$$

which is interpreted as meaning that learning F tells you nothing new about E.

Hempel's paradox provides an example. We are surprised by Hempel's argument because we think that the color of ravens is independent of the color of other birds. If we write this requirement into the model of figure 5.1, then $h_{ij} = p_i \times q_j$ and so

$$\text{prob}(E \,|\, G) = \frac{\text{prob}(E \text{ and } G)}{\text{prob}(G)} = \frac{p_1 q_1}{p_1 q_1 + p_1 q_2 + p_1 q_3} = q_1.$$

In this situation, sighting a pink flamingo therefore has no effect on the probability that all ravens are black.[11]

5.5.3 Updating on Zero-Probability Events

Since we are never allowed to divide by zero, the definition (5.4) for a conditional probability is meaningless when $\text{prob}(F) = 0$. In 1889, Joseph Bertrand (1889) offered some paradoxes that illustrate how one can be led astray by failing to notice that one is conditioning on an event with probability zero. The versions given here concern the problem of finding the probability that a randomly chosen chord to a circle is longer than its radius.

---

[11] In this analysis, p_1 is the probability that the bird which isn't a raven isn't black, and p_2 is the probability that it is. The probability that both ravens are black is $q_1 = h_{11} + h_{21}$, the probability that one raven is black is q_2, and the probability that no ravens are black is q_3.

One might first argue that the midpoint of a randomly chosen chord is equally likely to lie on any radius and that we therefore might as well confine our attention to one particular radius. Given that the midpoint of the chord lies on this radius, one might then argue that it is equally likely to be located at any point on the radius. With this set of assumptions one is led to the conclusion that the required probability is $\sqrt{3}/2 = 0.87$ (section 10.3). On the other hand, if one assumes that the midpoint of the chord is equally likely to be anywhere within the circle, then one is led to the conclusion that the required probability is $\frac{3}{4} = 0.75$ (section 10.3). Alternatively, we could assume that one endpoint of the chord is equally likely to be anywhere on the circumference of the circle, and that the second endpoint is equally likely to be anywhere else on the circumference, given that the first endpoint has been chosen. The required probability is then $\frac{2}{3} = 0.67$ (section 10.3).

In such cases, Kolmogorov (1950) pragmatically recommends considering a sequence of events F_n with $\text{prob}(F_n) > 0$ that converges on F. One can then seek to define $\text{prob}(E \mid F)$ as the limit of $\text{prob}(E \mid F_n)$. However, different answers can be obtained by considering different sequences of events F_n. In geometrical examples, it is easy to deceive oneself into thinking that there is a "right" sequence, but we can easily program a computer to generate any of our three answers by randomizing according to any of our three recipes. For example, to generate a probability of approximately 0.87, first break the circumference of the circle down into a large number of equally likely pixels. Whichever pixel is chosen determines a radius that can also be broken down into a large number of equally likely pixels, each of which can be regarded as the midpoint of an appropriate chord.

One may think that such considerations resemble the cogitations of medieval schoolmen on the number of angels that can dance on the point of a pin, but one can't escape conditioning on zero-probability events in game theory. Why doesn't Alice take Bob's queen? Because the result of her taking his queen is that he would checkmate her on his next move. Alice's decision not to take Bob's queen is therefore based on her conditioning on a counterfactual event—an event that won't happen. I believe that the failure of game theorists to find uncontroversial refinements of Nash equilibrium can largely be traced to their treating the modeling of counterfactuals in the same way that one is tempted to treat the notion of random choice in Kolmogorov's geometrical examples. There isn't some uniquely correct way of modeling counterfactuals. One must expand the set of assumptions one is making in order to find a framework within which counterfactual conditionals can sensibly be analyzed (Binmore 2007b, chapter 14).

5.6   Upper and Lower Probabilities

Upper and lower probabilities can't properly be said to be classical, but it is natural to introduce some of the ideas here because they are usually studied axiomatically. A whole book wouldn't be enough to cover all the approaches that have been proposed, so I shall be looking at just a small sample.[12]

In most accounts, upper and lower probabilities arise when Pandora doesn't know what probability to attach to an event, or doesn't believe that it is meaningful to say that the event has a probability at all. I separate these two possibilities into the case of ambiguity and the case of uncertainty, although we shall find that the two cases are sometimes dual versions of the same theory (section 6.4.3).

5.6.1   Inner and Outer Measure

After the Banach–Tarski paradox, it won't be surprising that we can't say much about nonmeasurable events in Kolmogorov's theory. But it is possible to assign an upper probability and a lower probability to any nonmeasurable event E by identifying these quantities with the outer and inner measure of E calculated from a given probability measure p.

We already saw how this is done in section 5.2.1 when discussing Lebesgue measure. The upper and lower probabilities will correspond to the outer and inner measures $\overline{m}$ and $\underline{m}$ defined by (5.1) and (5.2) when the collection C consists of all measurable sets.

In brief, the upper probability $\overline{p}(E)$ of E is taken to be the largest number that is no larger than the probabilities of all measurable sets that contain E. The lower probability $\underline{p}(E)$ of E is taken to be the smallest number no smaller than the probabilities of all measurable sets that are contained in E.[13] Since an event E is measurable in Kolmogorov's theory if and only if its upper and lower probabilities are equal, we have that

$$p(E) = \overline{p}(E) = \underline{p}(E), \qquad (5.6)$$

whenever E is measurable and so $p(E)$ is meaningful.

---

[12] Some references are Berger (1985), Breese and Fertig (1991), Chrisman (1995), DeGroot (1978), Dempster (1967), Earman (1992), Fertig and Breese (1990), Fine (1973, 1988), Fishburn (1970, 1982), Gardenfors and Sahlin (1982), Ghirardato and Marinacci (2002), Gilboa (2004), Giron and Rios (1980), Good (1983), Halpern and Fagin (1992), Huber (1980), Kyburg (1987), Levi (1980, 1986), Montesano and Giovannoni (1991), Pearl (1988), Seidenfeld (1993), Shafer (1976), Wakker (1989), Walley and Fine (1982), and Walley (1991).

[13] If F stands for a measurable set, then

$$\underline{p}(E) = \sup_{F \subseteq E} p(F), \qquad \overline{p}(E) = \inf_{E \subseteq F} p(F).$$

Hausdorff's paradox of the sphere provides a good example. If we identify the sphere with the surface of Mars, we can ask with what probability a meteor will fall on one of Hausdorff's three sets. But we won't get an answer if we model probability as Lebesgue measure on the sphere, because each of Hausdorff's sets then has a lower probability of zero and an upper probability of one (section 10.2).

Some examples of other researchers who take the line outlined here are Irving (Jack) Good (1983), Joe Halpern and Ron Fagin (1992), and Patrick Suppes (1974). John Sutton (2006) offers a particularly interesting application to economic questions.

5.6.2 Ambiguity

The use of *ambiguity* as an alternative or a substitute for *uncertainty* seems to originate with Daniel Ellsberg (1961), who made a number of important contributions to decision theory before he heroically blew the whistle on Richard Nixon's cynical attitude to the loss of American lives in the Vietnam War by leaking what became known as the Pentagon papers. The Ellsberg paradox remains a hot research topic nearly fifty years after he first proposed it.

The Ellsberg paradox. Figure 5.2 explains the paradox. There is no way of assigning probabilities to the three possible events that is consistent with the choices most people make.

The standard explanation is that the choices people make reflect an aversion to ambiguity. They know there is a probability that a black ball will be chosen, but this probability might be anything between 0 and $\frac{2}{3}$. When they choose J over K, they reveal a preference for winning with a probability that is certain to be $\frac{1}{3}$ rather than winning with a probability that might be anything in the range $[0, \frac{2}{3}]$. When they choose L over M, they reveal a preference for winning with a certain probability of $\frac{2}{3}$ to winning with a probability that might be anything in the range $[\frac{1}{3}, 1]$.

Bayesianism holds that rational folk are able to attach subjective probabilities to all events. Even if they only know that the frequency with which a black or a white ball will be chosen lies in a particular interval, they are still somehow able to attach a probability to each possible frequency. The behavior of laboratory subjects faced with the Ellsberg paradox is therefore dismissed as irrational. But I agree with David Schmeidler, who reportedly began his influential work on non-Bayesian decision theory with the observation that he saw nothing irrational in his preferring to settle issues by tossing a coin known to be fair, rather than tossing a coin about which nothing is known at all (see Gilboa and Schmeidler 2001).

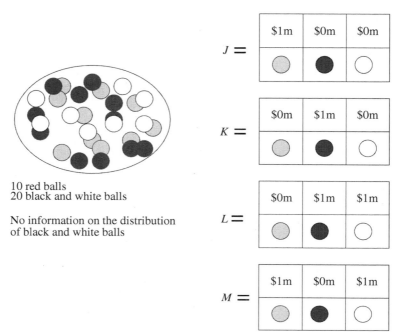

10 red balls
20 black and white balls

No information on the distribution
of black and white balls

Figure 5.2. Ellsberg paradox. A ball is drawn from an urn containing ten red balls and twenty black or white balls. No further information is available on how many balls are black or white. The acts *J*, *K*, *L*, and *M* represent various reward schedules depending on the color of the ball that is drawn. Most laboratory subjects choose *J* rather than *K* and *L* rather than *M*. However, if the probabilities of picking a red, black, or white ball are respectively p, q, and r, then the first choice implies that $p > q$ and the second choice implies that $q > p$.

5.6.3 Intervals of Probabilities

Ellsberg's paradox makes it natural to think in terms of making decisions when the states of the world don't come equipped with unambiguous probabilities, as in the Von Neumann and Morgenstern theory. What happens if we only know that the probability of a state lies in some range?

Quasi-Bayesian decision theory. One approach is to see what can be rescued from the Von Neumann and Morgenstern theory without adding any extra postulates. Giron and Rios (1980) undertook this task on the assumption that the class of probability measures that Pandora thinks possible is always convex. That is to say, if she thinks two probability measures over the set of states are possible, then she thinks that any probabilistic mixture of the two measures is also possible. The probabilities attached to any particular event will then lie in an interval.

If Pandora chooses a rather than b only when the expected utility of the act a exceeds that of b for *all* the probability measures she thinks possible, then we are led to an *incomplete* preference relation over the set of all her acts. In some cases, her criterion will leave the question of whether a or b should be chosen undecided.

In the spirit of revealed-preference theory, one can then seek to reverse the logic by asking what consistency postulates on an incomplete choice function ensure that Pandora will choose as though operating Giron and Rios's quasi-Bayesian theory. However, I shan't pursue this branch of revealed-preference theory any further. Walley (1991) is a good reference for the details.

The maximin criterion again. We really need a theory in which Pandora's choices reveal a complete preference relation over the set of acts. One approach is to follow up a suggestion that originates in the work of Abraham Wald (1950)—the founder of statistical decision theory— who proposed a more palatable version of the maximin criterion of section 3.7.

When all that we can say about the possible probability measures over a state space is that they lie in a convex set, Pandora might first work out the expected Von Neumann and Morgenstern utility $\mathcal{E}u(L(p))$ of all lotteries $L(p)$ for each probability measure p that she regards as possible. She can then compute $\underline{\mathcal{E}}u(L(p))$ as the largest number no larger than each $\mathcal{E}u(L(p))$ for all possible probability measures p. She can also compute $\overline{\mathcal{E}}u(L(p))$ as the smallest number no smaller than each $\mathcal{E}u(L(p))$ for all possible probability measures p. However, only the lower expectation is treated as relevant in this approach, since the suggestion is that Pandora should choose whichever feasible action maximizes $\underline{\mathcal{E}}u(L(p))$.

It can sometimes be confusing that the literature on statistical decision theory seldom talks about maximizing utility functions but expresses everything in terms of minimizing loss functions. What economists call the maximin criterion then becomes the minimax criterion. However, translating Walley's (1991) work on what he calls *previsions* into the language of economics, we find that lower expectations always satisfy

$$\underline{\mathcal{E}}(x + y) \geqslant \underline{\mathcal{E}}(x) + \underline{\mathcal{E}}(y) \quad \text{(superadditive)}, \tag{5.7}$$

$$\underline{\mathcal{E}}(\alpha x + \beta e) = \alpha\underline{\mathcal{E}}(x) + \beta \quad \text{(affinely homogeneous)}, \tag{5.8}$$

where x and y are vectors of Von Neumann and Morgenstern utilities in which Pandora gets x_n or y_n with probability p_n, the vector e satisfies $e_n = 1$ ($n = 1, 2, \ldots$), and $\alpha \geqslant 0$. We shall meet these conditions again, but the immediate point is that if Pandora were observed making choices as though maximizing a utility function with these properties, than we could deduce that she is behaving as though applying the maximin criterion in the presence of a convex set of possible probability measures.

The (finite) superadditivity of (5.7) is always a straightforward consequence of taking the minimum (or infimum) of any additive quantity. One similarly always obtains (finite) subadditivity by taking the maximum (or supremum) of any additive quantity. The condition (5.8) that I have clumsily called affine homogeneity follows straightforwardly from the linear properties of expectations.

Gilboa and Schmeidler (2004) go much deeper into these questions, but, as with quasi-Bayesian decision theory, I shan't pursue this line any further.

Nonadditive probabilities?   There is another raft of proposals in which upper and lower probabilities are defined directly from a convex set of probability measures in much the same way as upper and lower expectations are defined from orthodox expected utility.

The upper probability $\overline{p}(E)$ of an event E is taken to be the smallest number that is no smaller than all possible probabilities for E. The lower probability $\underline{p}(E)$ of E is taken to be the largest number no larger than all possible probabilities for E. The difference between these concepts and the outer and inner measures of section 5.6.1 is that the problem is no longer that we may have no way to assign E any probability at all, but that we may now have many ways of assigning a probability to E.

Lower probabilities defined in this way have the following properties:

1. *For any E, $\underline{p}(E) \geqslant 0$ with equality for the impossible event $E = \varnothing$.*

2. *For any E, $\underline{p}(E) \leqslant 1$ with equality for the certain event $E = B$.*

3. *If E and F have no state in common, then*

$$\underline{p}(E \text{ or } F) \geqslant \underline{p}(E) + \underline{p}(F) \quad \text{(superadditivity)} \qquad (5.9)$$

The finite superadditivity asserted in (5.9) could be replaced by countable superadditivity when the probabilities from which lower probabilities are deduced correspond to (countably additive) measures, but only finite additivity can be asserted when the probabilities from which they are deduced correspond to (finitely additive) charges.

Upper probabilities have the same properties as lower probabilities, except that superadditivity is replaced by subadditivity. To see why, note that

$$\overline{p}(E) = 1 - \underline{p}(\sim E)),$$

where $\sim E$ is the complement of E (which is the set of all states in B not in E).

Set functions that satisfy properties additional to the three given above for $\underline{p}$ are sometimes proposed as nonadditive substitutes for probabilities. David Schmeidler (2004) is a leading exponent of this school. One reason for seeking such a strengthening is that the three properties alone are not enough to imply that $\underline{p}$ is a lower probability corresponding to some convex set of possible probabilities. Imposing the following supermodularity condition (also called 2-monotonicity or convexity) on an abstract lower probability closes this gap. Whether the events E and F overlap or not, we require that:

$$\underline{p}(E \text{ or } F) + \underline{p}(E \text{ and } F) \geqslant \underline{p}(E) + \underline{p}(F) \quad \textit{(supermodularity)}.$$

Schmeidler defends supermodularity as an expression of ambiguity aversion, but other authors have alternative suggestions (Gilboa 2004).

This discussion of nonadditive probabilities only scratches the surface of an enormous literature in which the notion of a Choquet integral is prominent. However, I shall say no more on the subject, since my own theory takes me off in a different direction.

6
Frequency

6.1 Interpreting Classical Probability

Kolmogorov's classical theory of probability is entirely mathematical, and so says nothing whatever about the world. To apply the theory, we need to find an interpretation of the objects that appear in the theory that is consistent with whatever facts we are treating as given in whatever model of the world we choose to maintain.

The literature recognizes three major interpretations of probability:

- Objective probability
- Subjective probability
- Logical probability

Donald Gillies's (2000) excellent book *Philosophical Theories of Probability* surveys the literature very thoroughly from a perspective close to mine, and so I need only sketch the different interpretations.

Objective probability.   We assume that probabilities are objective when making decisions under risk (section 3.1).

Theories of objective probability assume the existence of physical entities that generate sequences of outcomes about which one can say nothing useful beyond recording the frequency with which each outcome occurs in a very large number of trials. The length of time before a uranium atom decays is a good example.

Roulette wheels are an artificial device for generating the long sequences of trials that are necessary to make objective probabilities meaningful. I am particularly interested in such artificial randomizing devices, because they are needed in game theory to implement mixed strategies.

Subjective probability.   We appeal to the theory of subjective probability when making decisions under uncertainty, although the circumstances under which this is legitimate remain controversial (section 3.1).

The subjective approach originates with the work of Frank Ramsey (1931) and Bruno de Finetti (1937). They invented what might be called a theory of revealed belief, in which Pandora's beliefs are deduced from her behavior—just as her preferences are deduced from her behavior in the later theory of revealed preference.

If Pandora's betting behavior satisfies certain consistency requirements, the theory of subjective probability argues that she will act as though she believes that each state of the world in B has a probability. These probabilities are said to be subjective, because Pandora's beliefs may be based on little or no objective data, and so there is no particular reason why Quentin's subjective probabilities should be the same as Pandora's.

The theory of subjective probability is generally thought to have been brought to a state of perfection in Leonard (Jimmie) Savage's (1951) much admired *Foundations of Statistics*. Chapter 7 describes how he succeeded in melding the ideas of Ramsey and de Finetti with those of Von Neumann and Morgenstern (section 7.2).

Bayes' rule figures large in applications of the theory of subjective probability. Suppose, for example, that Pandora is observing the play at a roulette table. If she assigns each possible probability measure over the 37 numbers a positive prior probability, then Bayesian updating will always result in her subjective probabilities converging on the objective probabilities associated with the roulette wheel.[1]

The emphasis on Bayes' rule seems excessive to classical statisticians. It was presumably for this reason that Ronald Fisher dismissively referred to Savage's synthesis as Bayesian decision theory. But the name stuck, and so people like me who see no problem in applying Savage's theory to the small worlds in which he intended it to be used can nowadays be called Bayesians without any pejorative overtones.

Logical probability. A logical or epistemic probability for a proposition is the rational degree of belief it derives from the available evidence. Some authors speak instead of the credence or level of credibility of the proposition.

The logical interpretation of probability is the most difficult of the three considered in this book, but Rudolf Carnap (1950) and Karl Popper (1963) nevertheless thought it worthwhile to devote time to exploring its possibilities. The young John Maynard Keynes (1921) sought to establish his intellectual reputation with a book on the subject, but he was only the first of many who have found the project too ambitious. I think we

---

[1] However, Diaconis and Freedman (1986) show that convergence on the true value of a parameter isn't always guaranteed when the number of possible outcomes is infinite, even when the prior distribution is restricted in an attempt to eliminate the problem.

are still so far from coming up with an adequate theory that I shall make no attempt to say anything systematic on the subject.

Not everyone agrees with this assessment. The prevailing orthodoxy in economics is Bayesianism, which I take to be the philosophical position that Bayesian decision theory always applies to all decision problems. In particular, it proceeds as though the subjective probabilities of Savage's theory can be reinterpreted as logical probabilities without any hassle. So its adherents are committed—whether they know it or not—to the claim that Bayes' rule solves the problem of scientific induction.

Outside economics, this kind of mindless Bayesianism is not so entrenched. One is faced instead with a spectrum of shades of opinion. I don't suppose it is really possible to identify all the 46,656 different kinds of Bayesian decision theory that Jack Good claimed to identify, but there is certainly a whole range of views between the painfully naive form of Bayesianism common in economics and the bare-bones version of Bayesian decision theory that I defend in this book.   ⤷ + essential

Let a thousand flowers bloom?   Which is the right interpretation of probability? Fisher was firmly of the opinion that only objective probability makes sense. De Finetti was equally determined that only subjective probability is meaningful.

I subscribe to Carnap's (1937) principle of tolerance. To insist either that there is a uniquely correct way to write a formal model of a particular intuitive concept or that there is a uniquely correct way to interpret a formal mathematical model is to venture into metaphysics. All three interpretations of probability therefore seem to me worthy of exploration. However, my own development of these interpretations doesn't treat them as conceptually independent.

I think the three interpretations are more usefully seen as a system of expanding circles in which, for example, the randomizing devices that generate the data which determine objective probabilities are regarded as entities within the world in which Pandora makes the judgments that determine her subjective probabilities. Subjectivists may find it tiresome that adopting this point of view necessitates some hard thinking about Richard von Mises' theory of randomizing devices, but that is the topic to which this chapter is devoted.

6.2   Randomizing Devices

Why do we need a theory of randomizing devices? In the case of a roulette wheel, for example, isn't it enough to follow the early probabilists by simply assuming that each number is equally likely? In the case of a uranium atom, why not simply assume that it is equally likely to decay in any fixed time period, given that it hasn't decayed already? There is a shallow answer, and what I regard as a deeper answer.

Osselots. When visiting India, I was taken to a palace of the Grand Mogul to see the giant marble board on which Akbar the Great played Parcheesi (Ludo) using beautiful maidens as pieces. Instead of dice, he threw six cowrie shells. If all six shells landed with their open parts upward, one could move a piece 25 squares—hence *parcheesi*, which is derived from the Hindi word for 25.

The Romans and Greeks liked to gamble too. Dice were not unknown, but the standard randomizing device was the osselot, which is the knee bone or heel bone of a sheep or goat. An osselot has six faces like a dice, but two of these are rounded, and so only four outcomes are possible when an osselot is thrown. Four osselots were usually thrown together. The *venus* throw required all four osselots to show a differently labeled face.

No two osselots are the same. The propensity of producing any particular outcome will therefore differ between osselots. Alfréd Rényi (1977) reports an experiment in which he threw an osselot one thousand times. The relative frequencies of each of the four faces were 0.408, 0.396, 0.091, and 0.105. A statistician would now be happy to estimate the probability of coming up with a *venus* in four independent throws of Rényi's osselot as 24 times the product of these four relative frequencies (24 is the number of ways that four different faces can arise in four throws). The answer is 0.0371, which explains why the ancients used the throw of a *venus* as the exemplar of an unlikely event.

The point of this excursion into ancient gambling practices is that one can't say anything accurate about the propensity of a particular osselot to fall on one of its faces without carrying out an experiment. It is true that, given a new osselot, one can appeal to experiments carried out on other osselots. For example, when estimating the probability of throwing a *venus* in the previous paragraph, it was assumed that Rényi's osselot was typical of ancient osselots. But who knows what conventions governed the choice of the osselots in ancient Rome? What were the regional variations? How well could an expert estimate the probability of a face on a new osselot simply by examining it very closely? By how much has selective breeding changed domestic sheep and goats over two thousand years? Contemplating such imponderables makes it clear that 0.0371 can't be treated as an *objective* estimate of the probability of throwing a *venus* in ancient times.

The necessity of doing empirical work with a particular osselot is plain, but the same is true of dice or roulette wheels if we are looking for real accuracy. It is true that modern dice are much more alike than osselots, but no two dice are exactly the same. Modeling dice as fair is therefore an idealization that is only justified up to a certain degree of accuracy. A million-dollar gambling scam in 2004 at London's Ritz Casino illustrates the point. Bets could be placed after the roulette wheel was spun, but it turns out that one can do much better than guess that the outcome will be random if one comes equipped with a modern computer and a laser scanner that observes how and when the croupier throws the ball onto the wheel.

Philosophers of the skeptical school will go further and even ask how we know that a pair of perfectly symmetric dice will roll *snake eyes* (two 1s) with probability $\frac{1}{36}$. Is it a synthetic a priori that a single perfectly symmetric dice will roll 1 with probability $\frac{1}{6}$?[2] Is it similarly a synthetic a priori that two dice rolled together are independent? When we ask the second question, are we saying that neither dice has any systematic effect on the other? If so, how are such considerations of cause and effect linked to Kolmogorov's definition of independence?

The purpose of all this questioning is to make it clear that David Hume (1975) had good reason to argue that our knowledge of all such matters is based only on experience. If a work of philosophy tries to convince us of something, Hume suggests that we ask ourselves two questions. Does it say something about the logical consequences of the way we define words or symbols? Does it say something that has been learned from actually observing the way the world works? If not, we are to commit it to the flames, for it can contain nothing but sophistry and illusion.

The Vienna circle took this advice to an extreme that modern empiricists now largely disavow. It was thought, for example, that all definitions must be operational—that one can't meaningfully postulate entities or relationships that aren't directly verifiable by experiment. Nowadays, it is more usual to argue that a theory stands or falls on the extent to which its testable predictions are verified, whether or not all the propositions of the theory from which the predictions are deduced are directly testable themselves. This is the position I take in seeking to refine Richard von Mises' theory, although he himself seems to have subscribed to the less relaxed philosophical prejudices of his time. However, von Mises' insistence that objective probability must be based in principle on experimental data remains the foundation of the approach.

---

[2] Immanuel Kant (1998) claimed that some contingent facts about the world are known to us without our needing to consult our experience. His leading example of such a synthetic a priori was that space is Euclidean.

Living in sin? Von Neumann apparently said that anyone who attempts to generate random numbers by deterministic means is living in a state of sin. In spite of the random number generators that inhabit our computers and the success of internet casinos advertising randomizing algorithms that successfully mimic real roulette wheels, this attitude still persists. Traditionalists reject all algorithmic substitutes for a roulette wheel on the grounds that only physical randomizing devices can generate numbers that are "truly" random. But what does random mean? What tests should we apply to tell whether a device is random or not?

Kolmogorov (1968) was not a traditionalist in this regard. He was even willing to suggest a definition of what it means to say that a finite sequence of *heads* and *tails* is random. In recent years, the idea of algorithmic randomness has been developed much further by Greg Chaitin (2001) and others. To paraphrase Cristian Calude (1998), this computational school seeks to make precise the idea that an algorithmically generated random sequence will fail all effective pattern-detecting tests. Randomness is therefore essentially identified with unpredictability.

This idea is important to me because of its implications in game theory. Suppose, for example, that Alice is to play Bob at Matching Pennies (figure 9.5). In this game, Alice and Bob each show a coin. Alice wins if both coins show the same face. Bob wins if they show different faces. Every child who has played Matching Pennies in the playground knows the solution. In a Nash equilibrium, Alice and Bob will each play *heads* and *tails* with equal probability. How is the randomizing done? The traditional answer is that you should toss your coin before showing it.

If Alice plays Bob at Matching Pennies one hundred times in a row, she might toss her coin one hundred times in advance of playing. But what if it always falls *heads*? After ten games, Bob might conjecture that she will always play *heads*. When his conjecture fails to be refuted after ten more games, he might then always play *tails* and so win all the time. Alice will then wish she had chosen her sequence of *heads* and *tails* to satisfy some criterion for randomness like Kolmogorov's.

A traditionalist will reply that an implicit assumption in rational game theory is that it is common knowledge that each player is rational. Bob will therefore know for sure that Alice chose her sequence randomly, and so her past play offers no genuine information about her future play. But I think that a theory that isn't stable in the face of small perturbations to its basic assumptions isn't very useful for applied work.

For this reason, I think it interesting to seek to apply von Mises' theory of objective probability not only to natural randomizing devices but also to the kind of artificial devices that Von Neumann jocularly described as sinful.

6.3   Richard von Mises

Richard von Mises was the younger brother of Ludwig von Mises, who is still honored as a leading member of the Austrian school of economics. However, Richard's espousal of the logical positivism of the Vienna circle put him at the opposite end of the philosophical spectrum to his brother.

His professional career was in engineering, especially aerodynamics. He flew as a test pilot for the Austro-Hungarian army in the First World War, but when the Nazis came to power, his Jewish ancestry made it prudent for him to take up a position in Istanbul. Eventually he was appointed to an engineering chair at Harvard in 1944. As with Von Neumann, his contributions to decision theory were therefore something of a sideline for him.

6.3.1   Collectives

In von Mises' (1957, p. 12) theory, a *collective* is a sequence of trials that differ only in certain prespecified attributes. For example, the throws of pairs of dice in a game of craps are a candidate for a collective in which the attribute of a single throw consists of the pips shown on the two dice in that throw. The peas grown by Gregor Mendel in his pioneering work on genetics are a candidate for a collective in which the attributes of a single pea are the seven traits to which he chose to pay attention.

The probability of a particular attribute of a single trial within a collective is defined to be the limiting value of its relative frequency as more and more trials are considered. For example, in the sequence of *heads* and *tails*

$$H, T, H, T, H, T, H, T, H, T, \ldots, \tag{6.1}$$

the attribute of being a *head* has probability $\frac{1}{2}$ because its sequence of relative frequencies as we consider more and more trials is

$$1, \frac{1}{2}, \frac{2}{3}, \frac{1}{2}, \frac{3}{5}, \frac{1}{2}, \frac{4}{7}, \frac{1}{2}, \frac{5}{9}, \frac{1}{2}, \ldots,$$

which converges to $\frac{1}{2}$.

However, we wouldn't say that the coin that generated the sequence (6.1) has a propensity of one half to generate a *head* at every trial. Rather, it has a propensity to generate a *head* for sure at even trials and a *tail* for sure at odd trials. But von Mises wants to say that the same probability is to apply to an attribute no matter where in the sequence the trial we are talking about occurs. He therefore insists that his collectives are random.

What does random mean? Von Mises says that a collective is random if its probability remains unchanged when we pass from the whole sequence to a certain class of subsequences. For example, the sequence (6.1) isn't random because the probability of the collective changes when we pass from the whole sequence of trials to the subsequence of odd trials.

Which class of subsequences do we need to consider when asking whether a collective is random? We can't consider all subsequences because the subsequence consisting of all *heads* will have probability one whatever the probability of the original sequence may be.

Von Mises (1957, p. 25) tells us that the question of whether or not a certain trial in the original sequence should be allowed to belong to the subsequence must be settled independently of the result of that trial or future trials. Nowadays we would say that the subsequence needs to be defined *recursively* in terms of the outcomes only of past trials. Following work by Abraham Wald (1938) and others, Alonzo Church (1940) confirmed that nearly all sequences are random in this sense.

6.3.2 Von Mises' Assumptions

Von Mises makes two assumptions about collectives:

1. *The frequencies of the attributes within a collective converge to fixed limits.*

2. *The trials within a collective are random, so it doesn't matter which trial you choose to look at.*

He argues at length that these assumptions should be regarded, like any other natural laws, as summarizing experimental data.

Gambling systems. Von Mises defends his assumptions with an evolutionary argument (section 1.6). He points out that Pandora could be exploited if she tried to implement a mixed strategy using a randomizing device that failed his requirements for a collective. The next section suggests that she should be more picky still, but von Mises' argument will still apply.

Suppose that the mixed strategy Alice is seeking to implement in Matching Pennies calls for *heads* to be played with probability $\frac{1}{2}$. She plays repeatedly using a device that generates *heads* and *tails* with an overall probability of $\frac{1}{2}$. However, she neglects to pay attention to von Mises' second requirement. The result is that her sequence of *heads* and *tails* incorporates a recursively defined subsequence in which the probability of *heads* is $\frac{2}{3}$. If Bob identifies this subsequence, he will play *tails* for sure whenever the next term of the subsequence is due. Alice will therefore end up losing on this subsequence.

Dawid and Vovk (1999) survey a substantial literature that interprets probability theory in such game-theoretic terms, but since my ultimate aim is to solve some game-theoretic problems by extending the scope of probability theory, I am unable to appeal to this elegant work.

Von Mises himself draws an analogy with perpetual-motion machines in denying the possibility of a gambling system that can consistently beat the bank at Monte Carlo. He then offers the survival of the casino industry as evidence of the empirical validity of his two requirements for roulette wheels. The survival of the insurance industry is similarly evidence for the fact that "all men insured before reaching the age of forty after complete medical examination and with the normal premium" is a collective if, for example, we take the relevant attributes to be whether or not an insured man dies in his forty-first year.

In such examples, von Mises repeatedly insists that his notion of an objective probability is meaningful only within the context of a properly specified collective. The same will be true of my attempt to refine his theory in the next section.

6.3.3 Conditional Probability

Section 5.5.1 emphasizes that Kolmogorov's definition of a conditional probability is no more than a definition. When textbooks offer a reason for using his definition, they nearly always take for granted that probabilities are to be interpreted as frequencies. Von Mises' theory of objective probability allows this explanation to be expressed in formal terms.

We need two collectives. A set F of attributes pertaining to the second collective determines a subsequence of trials in the first collective—those trials in which the outcome of the corresponding trial in the second collective lies in the set F. We then regard this subsequence as a new collective. If E is a set of attributes pertaining to the first collective, von Mises defines the conditional probability prob$(E \mid F)$ to be the limiting relative frequency of E in the new collective.

To see why Kolmogorov's formula (6.2) follows, we return to the base-rate fallacy of section 5.5. The set I is the event that someone in a given population has a particular disease. The set P is the event that a test for the disease has produced a positive result. The numbers shown in figure 6.1 are consistent with the maintained hypothesis that the test gets the answer wrong 1% of the time.

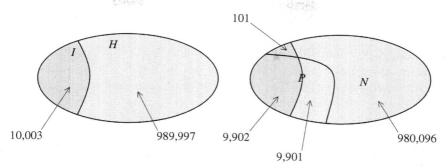

Figure 6.1. The base-rate fallacy. When laboratory subjects are given the actual numbers from an experiment, they make fewer mistakes (Gigerenzer 1996). The sample size in the figure is $1,000,000$. The number of ill people is 10,003. The number of people who test positive and are ill is $9,902$. The number who test positive is $9,902 + 9,901$. Thus $\#(P \text{ and } I) = 9,902$, $\#(P) = 19,803$, and $\#(S) = 1,000,000$. To realize that the probability of being ill having tested positive is about one half, one need only note that 9,902 is nearly equal to 9,901.

Nobody has any difficulty in seeing that the relative frequency with which people who tested positive are ill is

$$\frac{\#(P \text{ and } I)}{\#(P)} = \frac{\#(P \text{ and } I)}{\#(S)} \times \frac{\#(S)}{\#(P)},$$

where $\#(S)$ is the number of people in the whole sample S, $\#(P \text{ and } I)$ is the number of ill people who test positive, and $\#(P)$ is the number of people who test positive. If we let the sample size tend to infinity, the equation converges to

$$\text{prob}(I \,|\, P) = \frac{\text{prob}(P \text{ and } I)}{\text{prob}(P)}, \tag{6.2}$$

provided that the data conforms to von Mises' requirements for a collective and $\text{prob}(P) \neq 0$.

Independence. Von Mises' theory also offers the following objective definition of what it means to say that two collectives are independent. If we confine our attention to a subsequence of one collective obtained by looking only at trials of which the outcome is a particular attribute of the second collective, then the probability of the subsequence must be the same as the probability of the original collective. That is to say, knowing the outcome of the second collective tells us nothing more than we already knew about the prospective outcome of the first collective.

Countable additivity? The objective probabilities defined by von Mises' theory are finitely additive because the sum of two convergent sequences converges to the sum of their limits. However, as de Finetti (1974a,

p. 124) gleefully pointed out, von Mises' probabilities aren't necessarily countably additive.

I don't think that de Finetti was entitled to be so pleased, because it remains sensible to use countably additive models as simplifying approximations to objective probabilities in some contexts, just as it remains sensible to use continuous models in physics even though we now know that more can be explained in principle by treating space and time as though they were quantized (von Mises 1957, p. 37). On the other hand, de Finetti is right to insist that, within von Mises' formal theory, we can't hang on to countable additivity even if we allow the class of measurable sets to be very small—excluding perhaps all finite sets.

Consider, for example, the implications of defining the measure of a periodic set on the set $\mathbb{Z}$ of all integers to be its frequency. For each integer i, let P_i be a periodic set of integers with frequency $\frac{1}{3}\epsilon 2^{-|i|}$ that contains i. If we could extend our measure to be countably additive on some class of measurable sets, then the measure of $\mathbb{Z}$ would be less than the measure of the (countable) union of all P_i, which has been constructed to be no more than ϵ. But the measure of $\mathbb{Z}$ is 1.

6.4   Refining von Mises' Theory

My own view of von Mises' theory is that he is too ready to admit sequences as collectives. In particular, the relative frequency of an attribute of a collective should not simply converge on a limit p to be called its probability. The convergence should be *uniform* in a sense to be explained below.

6.4.1   Randomizing Boxes

In explaining why I think a new approach is useful, I shall stop talking about collectives and attributes to avoid confusion with von Mises' original theory. Instead of a collective, I shall speak of a *randomizing box* whose name or label is an infinite sequence of real numbers belonging to an *output set A*. The label is to be interpreted as an idealized history of experimentation with a black box responsible for generating that output.

The output set A is the analogue of von Mises' set of possible attributes. When the elements of the output set are probabilities, we can think of the randomizing box as an idealized implementation of a compound lottery. By reinterpreting *heads* and *tails* in such compound lotteries as $\mathcal{W}$ and $\mathcal{L}$ in Von Neumann and Morgenstern's theory, we can think of randomizing boxes as acts among which Pandora may have preferences (section 9.2.1).

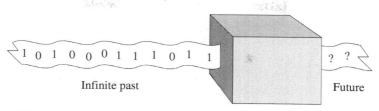

Figure 6.2. A randomizing box. The tape is to be envisaged as moving from right to left through the black box. When offered a new space, the box prints a 0 or a 1. It has printed an infinite sequence of such 0s and 1s in the past. This past history of the box serves as its label or name. What will be the next digit that emerges from the box?

In the simplest case, $A = \{0, 1\}$ is the set consisting of only 0 and 1. The randomizing box is then a generalization of a coin-tossing device, with the fall of a *head* corresponding to the term 1 in the sequence and the fall of a *tail* corresponding to the term 0. Figure 6.2 represents this case as a traditional black box that has printed out an infinite sequence of 0s and 1s in the past, and the question is what we choose to say now about the next digit that it will print.

Skeptics argue that we waste our time trying to find an a priori reason why the next output of a randomizing box should be one thing rather than another. Didn't David Hume teach us that we can't even give an a priori reason to believe that the sun will rise tomorrow? However, the project here isn't to find an a priori solution to the problem of scientific induction. As von Mises explained, the ultimate defense of his theory is experimental. Does it work in practice?

6.4.2   Uniform Convergence

When the attributes are *heads* and *tails*, von Mises requires of a collective that

$$f_n = \frac{x_1 + x_2 + \cdots + x_n}{n} \to p \quad \text{as } n \to \infty, \tag{6.3}$$

where $x_k = 1$ if the kth trial is a *head* and $x_k = 0$ if it is a *tail*. The number p is then said to be the probability of *heads* within the collective.

This convergence requirement is to be compared with the outcome of a sequence of Bernoulli trials in which an unfair coin weighted to come down *heads* with probability p is tossed repeatedly. The strong law of large numbers then says that $f_n \to p$ as $n \to \infty$ with probability one. It isn't impossible that every trial might be a *head* (so that $f_n \to 1$ as $n \to \infty$), but the whole class of sequences that fail to satisfy (6.3) has probability zero.

An elementary proof of a version of the strong law of large numbers for the case when $p = \frac{1}{2}$ is to be found in Billingsley (1986, pp. 5–11).

Following Émile Borel (1905), a real number x is first chosen from a uniform distribution on $[0, 1]$. The binary expansion of x is then taken to be the sequence of 0s and 1s whose limiting frequency is to be examined. A constructive argument is then given to show that the set of x for which (6.3) fails has Lebesgue measure zero.

This proof is easily adapted to show that the set of x for which

$$f_n(m) = \frac{x_{m+1} + x_{m+2} + \cdots + x_{m+n}}{n} \to p \quad \text{as } n \to \infty \qquad (6.4)$$

uniformly in m also has probability one.[3] There is, of course, a sense in which this result must hold, since it would be very odd if throwing away a finite number of terms at the beginning of a sequence of *independent* trials could make a difference to limiting properties of the sequence that hold with probability one.

This fact seems a good reason for replacing von Mises' requirement that $f_n \to p$ as $n \to \infty$ by the requirement that $f_n(m) \to p$ as $n \to \infty$ uniformly in m. The substitution has substantive consequences, but some mathematical preliminaries are necessary before they can be explained. A good reference for the mathematical details is Goffman and Pedrick (1965, pp. 62–68).

Extending linear functions. The set X of all bounded sequences x of real numbers is a vector space. In particular, given any two sequences x and y, the linear combination $\alpha x + \beta y$ is also a bounded sequence of real numbers. A subset Y of X that is similarly closed under linear combinations is a vector subspace of X. A linear function[4] $L : Y \to \mathbb{R}$ always satisfies

$$L(\alpha x + \beta y) = \alpha L(x) + \beta L(y).$$

We can always extend a linear function $L : Y \to \mathbb{R}$ defined on a subspace Y to the whole of a vector space X. Given any z not in Y, the set Z of all x that can be written in the form $y + yz$ is the smallest vector subspace of X that contains both Y and z. To extend L from Y to Z, take $L(z)$ to be anything you fancy and then define $L(x)$ for each $x = y + yz$ in Z by

$$L(x) = L(y) + yL(z).$$

---

[3] To say that $f_n(m) \to p$ as $n \to \infty$ means that for any $\epsilon > 0$ we can find $N(m)$ such that $|f_n(m) - p| < \epsilon$ for any $n > N(m)$. To say that $f_n(m) \to p$ as $n \to \infty$ uniformly in m means that the same value of N will suffice in the definition of convergence for all values of m.

[4] It is because we need to be precise about linearity at this stage that I was pedantic about calling affine functions affine in previous chapters.

In this way, we can extend L step by step until it is defined on the whole set X.[5]

We can actually extend L from Y to X while hanging on to some of its properties. Suppose that $\lambda : X \to \mathbb{R}$ is (positively) homogeneous and (finitely) subadditive. This means that, for all $\alpha \geqslant 0$, and for all x and y,

1. $\lambda(\alpha x) = \alpha \lambda(x)$;
2. $\lambda(x + y) \leqslant \lambda(x) + \lambda(y)$.

The famous (but not very difficult) Hahn–Banach theorem says that if $L(y) \leqslant \lambda(y)$ for all y in the subspace Y, then we can extend L as a linear function to the whole of X so that $L(x) \leqslant \lambda(x)$ for all x in X.

We noted one application in chapter 5. Lebesgue measure can be extended from the measurable sets on the circle as a finitely additive function defined on *all* sets on the circle. However, the application that we need in this chapter concerns the existence of Banach limits.

Extending probability? We now return to randomizing boxes in which the output set A consists of all real numbers in $[0, 1]$. Each such number will be interpreted as the probability of generating a *head* in a lottery that follows.

We are therefore talking about compound randomizing boxes that might implement the compound lotteries of section 3.4. However, in doing so we are getting ahead of ourselves, because we haven't yet decided exactly what is to count as a randomizing box, nor whether such randomizing boxes will serve to implement the kind of lotteries we have considered in the past. In pursuing these questions, von Mises' second requirement on randomness will be put aside for the moment so that we can focus on his first requirement. Why do I think this needs to be strengthened by appending a uniformity condition?

Suppose a randomizing box is labeled with a sequence x of real numbers x_n that represents its history of outputs. When can we reasonably attach a probability $p(x)$ to this sequence that can be interpreted as the probability of finally ending up with a *head* on the next occasion that the randomizing agent offers us an output?

To be interpreted as a probability, we need the function $p : X \to \mathbb{R}$ from the vector space of all bounded sequences to the set of real numbers to satisfy certain properties. It must be linear, so that lotteries compound according to the regular laws of probability. We then insist on the following minimal requirements:

---

[5] To justify this claim, it is necessary to appeal to Zorn's lemma (actually proved by Hausdorff) or some other principle that relies on the axiom of choice.

1. $p(x) \geqslant 0$ when $x_n \geqslant 0$ for all $n = 1, 2, \ldots$;

2. $p(e) = 1$ when $e_n = 1$ for all $n = 1, 2, \ldots$;

3. $p(x) = p(y)$ when y is obtained by throwing away a finite number of terms from x.

It is easy to locate a linear function p that satisfies these requirements on some vector subspace of X. We need only set $p(x)$ equal to the limit of any sequence x that lies in the subspace Y of convergent sequences. Is there an unambiguous way to extend p to a larger and more interesting subspace? Fortunately, this question was answered by a self-taught mathematical prodigy called Stefan Banach.

We can extend p from Y to the whole of X in many ways. Banach identified the smallest vector subspace Z on which all of these extensions agree. The sequences in the subspace Z are said to be "almost convergent." The value of $p(z)$ when z is an almost convergent sequence is called its Banach limit.

If the probability function p is to be extended unambiguously, we must therefore restrict it to the almost convergent sequences. This requirement entails appending our uniformity requirement (6.4) to von Mises' assumption that the relative frequency of a sequence x should converge to $p(x)$.

A finitely additive measure on the natural numbers. What intuition can be mustered for denying that a randomizing box necessarily has probability $\frac{2}{3}$ of generating a *head*, when its past history consists of a sequence y of *heads* and *tails* in which the relative frequency of *heads* converges to $\frac{2}{3}$? I think Banach teaches us that the distribution of the terms of y may be too wild for it to be meaningful to attach probabilities to its outputs. Von Mises' theory therefore leaves us in an uncertain world rather than a risky world.

Waiting for a bus is a possible analogy. With the uniformity condition, Pandora is guaranteed that the frequency with which buses arrive will be close to the advertised rate, within a long enough but finite period. Without the uniformity condition, the bus frequency may balance out at infinity, but what good is that to anybody?

We can flesh out the bones of this intuition with the following construction. Given a set S of natural numbers, define a sequence $x(S)$ by making its nth term equal to 1 if n belongs to S and 0 if it doesn't. The function μ defined on certain subsets of the natural numbers by

$$\mu(S) = p(x(S)) \qquad (6.5)$$

is then finitely additive. Roughly speaking, the definition says that the number of elements of S in any interval of length ℓ can be made as close as we like to $\mu(S) \times \ell$ by making ℓ sufficiently large.

Many authors would call μ a finitely additive measure, although it isn't countably additive, and so it isn't properly a measure at all. In section 5.3.1, we agreed that functions like μ would be called charges. The domain of μ can't be all subsets of natural numbers, because p is only defined on sequences in Z. The sets on which μ is defined should perhaps be called chargeable, but I am going to call them measurable, although one can only count on finite unions of measurable sets being measurable when talking about charges.

The charge μ treats every natural number equally. We have seen that there is no point in looking for a proper measure that does the same (section 6.3). The best that can be said for μ in this regard is that it is countably superadditive.

Returning to a randomizing box that fails the uniformity test, we can now observe that the set of natural numbers corresponding to *heads* in the sequence y isn't measurable according to μ. People who won't wait for buses that arrive when y produces a *head* can therefore be regarded as uncertainty averse, because they don't like events being nonmeasurable.

6.4.3 Objective Upper and Lower Probabilities?

What do we do when a randomizing box doesn't have a probability? What action should Alice take, for example, if Bob deliberately uses such a randomizing box when choosing his strategies in a game they are playing together?

Within von Mises' theory. Popper (1959) considered the possibility that the sequence f of relative frequencies given by (6.3) might not converge. His response was to draw attention to the cluster points of f, which he called "middle frequencies."[6] Walley and Fine (1982) explore this point of view in considerable detail.

A cluster point c has the property that we can find a subsequence of f that converges on c. As figure 6.3 indicates, the relative frequency f_n gets arbitrarily close to any cluster point c for an infinite number of values of n.

When x is a sequence of 0s and 1s, a Bayesian might hypothesize that a 1 occurs independently at each trial with probability p. If this were true, then updating by Bayes' rule from a prior probability measure

---

[6] Cluster points are also called points of accumulation or limit points.

that attaches positive probability to both 0 and 1 generates a posterior probability measure that eventually puts all its weight on the correct value of p. But when we do the same with a sequence x for which the associated sequence f of relative frequencies doesn't converge, then the posterior probability measure oscillates through all the cluster points of f, sometimes putting nearly all of its weight on one cluster point and sometimes on another.

This discussion suggests that a putative collective whose sequence of relative frequencies doesn't converge might be treated as a problem in ambiguity, with the interval I of figure 6.3 taking on the role of the convex set of probability measures among which Pandora is unable to make a selection (section 5.6.3).[7] The largest cluster point can then be taken to be the upper probability $\overline{q}(x)$ of getting a *head*, and the smallest cluster point can be taken to be the lower probability $\underline{q}(x)$. In mathematical notation:

$$\overline{q}(x) = \limsup_{n \to \infty} \frac{x_1 + x_2 + \cdots + x_n}{n},$$
$$\underline{q}(x) = \liminf_{n \to \infty} \frac{x_1 + x_2 + \cdots + x_n}{n}.$$

In the case when I contains only a single point because $\overline{q}(x) = \underline{q}(x)$, the sequence x is a candidate for a von Mises collective because it satisfies his first requirement (6.3). Its von Mises probability $q(x)$ is the common value of its upper and lower probabilities.

Refined upper and lower probabilities.   For any $\epsilon > 0$, we can find an N such that for any $n > N$

$$\underline{q}(x) - \epsilon < \frac{x_1 + x_2 + \cdots + x_n}{n} < \overline{q}(x) + \epsilon. \tag{6.6}$$

In fact, $\overline{q}(x)$ and $\underline{q}(x)$ are respectively the smallest and the largest numbers for which this assertion is true.

My theory of randomizing boxes requires that (6.6) be strengthened. For $\overline{p}(x)$ and $\underline{p}(x)$ to be upper and lower probabilities, we need them to be respectively the smallest and largest numbers for which the following uniformity condition holds.

For any $\epsilon > 0$, we can find an N such that for any $n > N$

$$\underline{p}(x) - \epsilon < \frac{x_{m+1} + x_{m+2} + \cdots + x_{m+n}}{n} < \overline{p}(x) + \epsilon \tag{6.7}$$

for all values of m.

---

[7] The set of cluster points of a sequence need not be convex in general, but if x is a bounded sequence, then the set of all cluster points of f is necessarily convex.

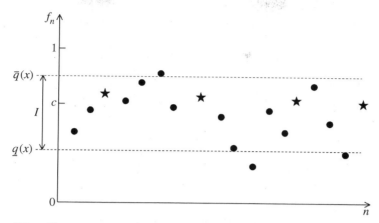

Figure 6.3. Cluster points. The figure shows a divergent sequence f of relative frequencies derived from a sequence x whose terms lie between 0 and 1. The starred terms of f are a subsequence that converges to the cluster point c. The interval $I = [\underline{q}(x), \overline{q}(x)]$ is the set of all such cluster points.

Because (6.7) is a stronger condition than (6.6),

$$\underline{p}(x) \leqslant \underline{q}(x) \leqslant \overline{q}(x) \leqslant \overline{p}(x). \tag{6.8}$$

Notice that a sequence x may have $\underline{q}(x) = \overline{q}(x)$ and so have a probability in the sense of von Mises, but nevertheless not have a probability in the refined sense, because $\underline{p}(x) < \overline{p}(x)$.

Properties of upper and lower probabilities. Recall that x can be interpreted as a sequence of Von Neumann and Morgenstern utilities. It is therefore unsurprising that $\underline{p}(x)$ behaves like a lower expectation (section 5.6.3):

$$\underline{p}(x + y) \geqslant \underline{p}(x) + \underline{p}(y) \quad \text{(superadditive),} \tag{6.9}$$

$$\underline{p}(\alpha x + \beta e) = \alpha \underline{p}(x) + \beta \quad \text{(affinely homogeneous),} \tag{6.10}$$

where $\alpha \geqslant 0$ and e is the sequence whose terms are all 1.

Analogous properties for $\overline{p}(x)$ follow (with superadditivity replaced by subadditivity), because $\overline{p}(e - x) = -\underline{p}(x - e) = -\underline{p}(x) + 1$, and so

$$\overline{p}(x) + \underline{p}(y) = 1 \quad \text{when } x + y = e. \tag{6.11}$$

Uncertainty. Section 5.6 briefly examined two ways of thinking about upper and lower probabilities. The first of these treated them as outer and inner measures (section 5.6.1).

Recall that (6.5) defines a charge (or finitely additive measure) on the set of natural numbers (section 6.4.2). For any set T of natural numbers, we construct a sequence $x(T)$ whose nth term is 1 if n lies in T and 0 if it doesn't. If the sequence $x(T)$ has a probability $p(x(T))$ in our refined sense, we say that the set T is measurable (or chargeable) and that its charge is $\mu(T) = p(x(T))$.

If S is nonmeasurable in this sense, we can define its inner charge $\underline{\mu}(S)$ to be the largest number no larger than every $\mu(T)$ for which T is a measurable subset of S. Similarly, we can define its outer charge $\overline{\mu}(S)$ to be the smallest number no smaller than every $\mu(T)$ for which T is a measurable superset of S.

Do our definitions of $\overline{\mu}(S)$ and $\underline{\mu}(S)$ accord with those of $\overline{p}(x(S))$ and $\underline{p}(x(S))$? In particular, is it true that $\overline{\mu}(S) = \overline{p}(x(S))$? To see that the answer is *yes*, we need to return to (6.7).

We can approximate $\overline{\mu}(S)$ as closely as we like by the charge $\mu(T)$ of a measurable superset T of S. In the case when T is measurable, our definitions say that the number of elements of T in any interval of length ℓ can be made as close as we like to $\mu(T) \times \ell$ by making ℓ sufficiently large. Our problem in connecting this fact with the definition of $\overline{\mu}(S)$ is that (6.7) only tells us that the measurability requirement for the set S itself holds for an infinity of intervals of length ℓ, but not necessarily for all such intervals. However, we can replace 0s by 1s in the sequence $x(S)$ until our new sequence $x(T)$ satisfies the measurability requirement for all intervals. The resulting measurable superset T of S then supplies the necessary connection between $\overline{\mu}(S)$ and $\overline{p}(x(S))$.

Ambiguity.   In the refined version of von Mises' theory, the uncertainty approach to upper and lower probabilities is mathematically dual to the ambiguity approach (section 5.6). There is therefore an important sense in which it doesn't matter which approach we take. Personally, I find the ambiguity approach intuitively easier to handle.

The theory of Banach limits (section 6.4.2) depends heavily on the properties of $\overline{p}(x)$ and $\underline{p}(x)$. Any Banach limit ℓ defined on the space of all bounded sequences is first shown to satisfy

$$\underline{p}(x) \leqslant \ell(x) \leqslant \overline{p}(x). \tag{6.12}$$

So all Banach limits are equal when $\underline{p}(x) = \overline{p}(x)$. It follows that there can be no ambiguity about how to define the probability $p(x)$ that a randomizing box will generate a *head* when x is almost convergent (section 6.4.2). We have no choice but to take $p(x) = \underline{p}(x) = \overline{p}(x)$.

But suppose that z isn't almost convergent, so that $\underline{p}(z) < \overline{p}(z)$. We can then extend p as a Banach limit from the space of almost convergent sequences so as to make $p(z)$ anything we like in the range:

$$I = [\underline{p}(z), \overline{p}(z)].$$

Simply take the function λ in the Hahn–Banach theorem to be $\overline{p}(x)$ (section 10.4).

This fact links my approach with the ideas on ambiguity briefly surveyed in section 5.6.3. If the set of all possible probability functions is identified with the set of all Banach limits, then $\overline{p}(z)$ is the maximum of all possible probabilities $p(z)$, and $\underline{p}(z)$ is the minimum.

Alternative theories? Someone attached to von Mises' original theory may reasonably ask why we shouldn't append additional requirements to the three conditions for a Banach limit given in section 6.4.2, thereby characterizing what might be called a super Banach limit. The space on which such a super Banach limit is defined will be larger than the set of all almost convergent sequences, and so more sequences will be assigned a probability.

For example, we could ask that von Mises' first requirement be satisfied, so that any x whose sequence of relative frequencies f converges to a limit p is assigned the value $q(x) = p$. The Hahn–Banach theorem with $\lambda = \overline{q}$ can then be used to extend q to the set of all bounded sequences. The upper and lower probabilities of x would then be $\overline{q}(x)$ and $\underline{q}(x)$.[8] But why not go further and require that $p(x) = p$ whenever x is (C, k) summable to p? Why not allow some even wider class of summability methods for assigning a generalized limit to a divergent sequence (Hardy 1956)?

I don't think that there is a one-and-only correct model. My refinement of von Mises' theory is simply the version that carries the least baggage.

6.5   Totally Muddling Boxes

Von Mises second requirement is that a collective be random. Recall that his definition requires that a collective with probability q should have the property that every effectively determined subsequence should also have probability q.

If von Mises had permitted collectives with upper and lower probabilities $\overline{q}$ and $\underline{q}$, he would presumably have asked that every effectively determined subsequence should also have upper and lower probabilities $\overline{q}$ and $\underline{q}$.[9] I don't know whether Alonzo Church's (1940) vindication of von Mises' second requirement would extend to this case, but even if it

---

[8] The argument of section 10.4 works equally well with p replaced by q.

[9] Karl Popper (1959, appendix IV) offers an alternative to von Mises' definition of randomness that allows for differing upper and lower probabilities, but it doesn't seem adequate to me.

did, it wouldn't be of any direct help in dealing with the randomizing boxes introduced in section 6.4.1.

Randomizing boxes diverge from collectives not just in satisfying a uniformity requirement, but in being structured differently (section 6.4). A randomizing box isn't envisaged as an ongoing process in which we seek to predict the probability that the next trial will be a *head*. The label x on a randomizing box is a full description of a countably infinite number of trials that have *already* been carried out.

Without changing anything that matters, one could therefore envisage the tape of figure 6.2 as being reversed, so that the first term of the sequence x lies infinitely far to the left of the box.[10] The digit about to be printed then has no immediate predecessor. No algorithm that examines only a finite number of terms of x can therefore say anything relevant about the value of $p(x)$, which depends only on the limiting properties of x. A randomizing requirement is therefore redundant.

This approach sweeps the randomness problem under the carpet by exploiting the fact that the label x of a randomizing box is idealized to be an *infinite* sequence. However, any attempt to produce a finite computer program that approximates the properties of a randomizing box wouldn't be able to avoid ensuring that the next digit to be printed can't ever be predicted in an appropriate sense by examining the digits already printed (Stecher and Dickhaut 2008).

Muddling boxes. A muddling box will be understood to be a randomizing box that offers Pandora no grounds for taking into account any of its properties other than the values of its upper and lower probabilities. For example, if it were meaningful to say that the terms of its sequence f of relative frequencies were only occasionally close to its lower probability, we would want to attach more importance to its upper probability. The weight we attached to the upper probability would then be an additional factor that would need to be taken into account when evaluating the randomizing box.

My suggestion for coping with this problem uses the explicit formulas:

$$\overline{p}(x) = \inf \left\{ \limsup_{n \to \infty} \frac{x_{m_1+n} + x_{m_2+n} + \cdots + x_{m_k+n}}{k} \right\},$$

$$\underline{p}(x) = \sup \left\{ \liminf_{n \to \infty} \frac{x_{m_1+n} + x_{m_2+n} + \cdots + x_{m_k+n}}{k} \right\},$$

in which formulas the infimum and supremum range over all finite sets $\{m_1, m_2, \ldots, m_k\}$ of natural numbers (section 10.5).

---

[10] Thereby inverting Wittgenstein's philosophical joke about the man he heard finishing a backward recital of the decimal expansion of π by saying ... 95141.3.

In the case of the sequence x of (6.1) in which *heads* and *tails* alternate, we have that $\overline{p}(x) = \underline{p}(x) = \frac{1}{2}$ (counting a *tail* as 0 and a *head* as 1). However, taking $k = 2$, $m_1 = 1$, and $m_2 = 3$,

$$\limsup_{n \to \infty} \frac{x_{1+n} + x_{3+n}}{2} = 1, \qquad \liminf_{n \to \infty} \frac{x_{1+n} + x_{3+n}}{2} = 0.$$

Looking at different sets $\{m_1, m_2, \ldots, m_k\}$ of natural numbers is therefore a way of eliciting the structure of a randomizing box.

The definition I propose for a *muddling box* is that for any $\epsilon > 0$,

$$\limsup_{n \to \infty} \frac{x_{m_1+n} + x_{m_2+n} + \cdots + x_{m_k+n}}{k} < \overline{p}(x) + \epsilon,$$

$$\liminf_{n \to \infty} \frac{x_{m_1+n} + x_{m_2+n} + \cdots + x_{m_k+n}}{k} > \underline{p}(x) - \epsilon,$$

for all finite sets $\{m_1, m_2, \ldots, m_k\}$ of natural numbers whose cardinality k is sufficiently large. A *totally muddling* box is a muddling box for which $\overline{p}(x) = 1$ and $\underline{p}(x) = 0$.

7

Bayesian Decision Theory

7.1 Subjective Probability

This chapter studies the theory of subjective probabilities invented independently by Frank Ramsey (1931) and Bruno de Finetti (1937). The version of the theory developed by Leonard Savage (1951) in his famous *Foundations of Statistics* is now universally called Bayesian decision theory. My own simplification of his theory in section 7.2 differs from the usual textbook accounts, but the end product will be no less orthodox than my simplification of Von Neumann and Morgenstern's theory of expected utility in section 3.4.

7.1.1 Small Worlds

We last visited the theory of revealed preference when developing a version of Von Neumann and Morgenstern's theory of expected utility (section 3.4). But Pandora must now face events for which the objective probabilities we associate with roulette wheels are unavailable. The archetypal example consists of betting at a race track.

We can't sensibly talk about the frequency with which *Punter's Folly* will win next year's Kentucky Derby because that particular race will only be run once. Pandora may have all kinds of more diffuse information about *Punter's Folly* and the other horses, but without an adequate theory of logical probability we don't know how to convert such information into an unambiguous probability. But people nevertheless bet on horses. Bookmakers bet so successfully that they sometimes get very rich.

Bayesian decision theory assumes that Pandora makes stable and consistent choices in the presence of events that lack objective probabilities (section 1.5). It characterizes her choice behavior in terms of both a Von Neumann and Morgenstern utility function u defined on her set C of consequences and a subjective probability measure p defined on her set B of states of the world. It is shown that she chooses from any feasible set A as though maximizing the expected value of her utility function u computed with respect to her subjective probability measure p.

When is it sensible to be consistent in the sense required to obtain such a result? It certainly can't be rational always to be consistent. Pandora should change her mind if she receives data that refutes a premise she has been taking for granted. For example, at one time nobody questioned that the Earth is flat, but now we think it is round. Savage (1951, p. 16) therefore restricted the application of his theory to what he called *small worlds* in which it makes sense to insist on consistency. He describes the idea that one can use his theory in any world whatever as "utterly ridiculous" and "preposterous."[1]

According to Savage, a small world is one within which it is always possible to "look before you leap." Pandora can then take account *in advance* of the impact that all conceivable future pieces of information might have on the underlying model that determines her subjective beliefs. Any mistakes built into her original model that might be revealed in the future will then *already* have been corrected, so that no possibility remains of any unpleasant surprises.

In a large world, the possibility of an unpleasant surprise that reveals some consideration overlooked in Pandora's original model can't be discounted. As Savage puts it, in a large world, Pandora can only "cross certain bridges when they are reached." Knee-jerk consistency is then no virtue. Someone who insists on acting consistently come what may is just someone who obstinately refuses to admit the possibility of error. In brief, Savage agrees with Ralph Waldo Emerson that foolish consistency is the hobgoblin of small minds. Only when our small minds are encased in a small world does he regard consistency as an unqualified virtue.

Although he is often cited as though he were the father of Bayesianism, Savage therefore disavowed this creed before it was even born. He didn't believe that Bayesian decision theory is the appropriate tool for all decision problems. He didn't think that Bayes' rule is the solution to the problem of scientific induction. And we shall see later that he didn't believe that rationality somehow endows us with prior probability measures with which to start off the process of Bayesian updating.

7.2 Savage's Theory

We start with the assumption that Pandora's choices reveal a stable and consistent preference relation over the set of all acts (section 1.5.2). In Von Neumann and Morgenstern's theory of expected utility, the events

---

[1] I am not sure how to reconcile these remarks with what (Savage 1951, p. 86) says later about "microcosms."

in the set B of all possible states come equipped with objective probabilities, so that we can identify an act with a lottery (section 3.4). But what if acts don't come equipped with objective probabilities, as at the race track? Instead of lotteries, we than have to deal with *gambles* of the form

$$G = \begin{array}{|c|c|c|c|c|} \hline G_1 & G_2 & G_3 & \cdots & G_n \\ \hline E_1 & E_2 & E_3 & \cdots & E_n \\ \hline \end{array} \qquad (7.1)$$

in which the events E_i need not have objective probabilities. The prizes G_i may themselves be acts like G, but it will be assumed that such nesting of acts doesn't go on for ever, so that Pandora eventually ends up with one of the final outcomes $\mathcal{P}_i$ in the set C of consequences after a finite number of steps.

The next postulate generalizes postulate 4, which says that Pandora doesn't care how compound acts are structured. She only cares about what prize she eventually gets from an act (section 3.7.1).

Postulate 7.   Pandora is indifferent between all acts that deliver the same consequences in the same states of the world.

If Pandora makes choices in a sufficiently consistent manner, Savage (1951) showed that we can assign *subjective* probabilities $p(E_i)$ to the events E_i in the gamble (7.1) in a way that makes Pandora indifferent between G and the lottery

$$L = \begin{array}{|c|c|c|c|c|} \hline G_1 & G_2 & G_3 & \cdots & G_n \\ \hline p(E_1) & p(E_2) & p(E_3) & \cdots & p(E_n) \\ \hline \end{array} \qquad (7.2)$$

All of Von Neumann and Morgenstern's postulates for lotteries are to be retained, and so it follows that Pandora has a Von Neumann and Morgenstern utility function U defined over the set $\aleph$ of all acts.[2]

In the case when the act G_i delivers the prize $\mathcal{P}_i$ for sure, Pandora will therefore choose whatever gamble G in her feasible set A maximizes

$$U(G) = p(E_1)u(\mathcal{P}_1) + p(E_2)u(\mathcal{P}_2) + \cdots + p(E_n)u(\mathcal{P}_n), \qquad (7.3)$$

where u is the restriction of Pandora's Von Neumann and Morgenstern utility to the set C of consequences, and p is her subjective probability measure defined on the set B of states of the world.

---

[2] We use the notation U for a Von Neumann and Morgenstern utility function with domain $\aleph$, and the notation u for its restriction to the set C of consequences (section 3.6.1).

$$a = \begin{array}{|c|c|} \hline \mathcal{L} & \mathcal{P} \\ \hline {\sim}E & E \\ \hline \end{array} \qquad\qquad b = \begin{array}{|c|c|c|} \hline \mathcal{L} & \mathcal{P} & \mathcal{Q} \\ \hline {\sim}(E \cup F) & E & F \\ \hline \end{array}$$

Figure 7.1. Two simple gambles. The outcome $\mathcal{P}$ in the gamble a on the left might be Pandora's winnings in the event E that *Punter's Folly* wins the Kentucky Derby. Since the prize $\mathcal{L}$ that she gets if *Punter's Folly* loses is her worst possible outcome, she must recklessly have bet everything she can beg or borrow on its winning.

The formula (7.3) makes manifest the separation between beliefs and preferences implicit in Savage's theory. However, we have yet to examine how Savage built this separation into his theory. David Kreps's (1988) excellent *Notes on the Theory of Choice* offers a version of Savage's own story, but I follow Anscombe and Aumann (1963) in taking a short cut.

Roulette wheels exist! Savage was unwilling to write objective randomizing devices like roulette wheels into his theory. He therefore proposed axioms that allow the construction of what one might call subjective randomizing devices. I think we find these axioms reasonable only because we know that objective randomizing devices actually do exist in real life. So why not postulate their existence up front?

Postulate 8. Enough events come equipped with objective probabilities that Pandora's set of acts includes all possible lotteries, including lotteries in which the prizes are themselves acts.

Postulate 8 allows us to appeal to Von Neumann and Morgenstern's postulates, and hence justify the existence of a Von Neumann and Morgenstern utility function U defined over Pandora's set $\aleph$ of acts (including the acts that make particular consequences in C certain). It is simplest to normalize U so that $U(\mathcal{L}) = 0$ and $U(\mathcal{W}) = 1$, where $\mathcal{L}$ and $\mathcal{W}$ are respectively the worst and best outcomes in Pandora's set C of consequences.

Having modeled Pandora's preferences, the next step is to ask what extra consistency requirements are needed to model her attitude to gambles in terms of subjective probabilities. For the moment, we look only at gambles a of the kind shown on the left in figure 7.1. In such gambles, Pandora gets the prize $\mathcal{P}$ if the event E occurs and the worst possible prize $\mathcal{L}$ if it doesn't. (The complement of E is denoted by $\sim E$.)

Write the Von Neumann and Morgenstern utility of the gamble a as

$$U(a) = V_E(\mathcal{P}),$$

where the dependence on $\mathcal{L}$ has been suppressed because this prize will be held fixed all the time. In the case when $E = B$ we recover the Von Neumann and Morgenstern utility function $u : C \to \mathbb{R}$ of section 3.6. That is to say,

$$u(\mathcal{P}) = V_B(\mathcal{P}). \tag{7.4}$$

The following simplifying assumptions don't have enough content to be counted as postulates. They exclude the possibility that there might be events of probability zero that aren't impossible, but nothing important hinges on this exclusion. Whenever $\varnothing \subset E \subset B$ and $\mathcal{L} \prec \mathcal{P} \prec \mathcal{W}$,

$$0 = V_E(\mathcal{L}) \ < V_E(\mathcal{P}) < V_E(\mathcal{W}) < V_B(\mathcal{W}) = 1,$$
$$0 = V_\varnothing(\mathcal{P}) < V_E(\mathcal{P}) < V_B(\mathcal{P}) \ < V_B(\mathcal{W}) = 1,$$

where the empty set $\varnothing$ represents impossibility and the whole state space B represents certainty.

Separating preferences and beliefs. The next postulate captures the part of Aesop's principle that requires Pandora to separate her preferences from her beliefs (section 1.4). We thereby exclude such wishful thinking as: I don't believe that because I don't like what it implies. Or, equally irrationally: I don't want that because I'm not likely to get it.

As with postulate 3, it needs to be understood that the underlying lottery is independent of the events that later determine what Pandora wins (section 3.4.1).

Postulate 9. If E isn't the empty event, the preferences revealed by Pandora's choices over lotteries of the form

$\mathcal{L}$	$\mathcal{P}_1$	$\mathcal{L}$	$\mathcal{P}_2$	$\mathcal{L}$	$\mathcal{P}_3$			$\mathcal{L}$	$\mathcal{P}_n$
$\sim E$	E	$\sim E$	E	$\sim E$	E	$\cdots$		$\sim E$	E
q_1		q_2		q_3		$\cdots$		q_n	

remain unchanged if E is replaced everywhere by any other nonempty event.

Postulate 9 won't be satisfied unless the underlying decision problem has been constructed suitably. Consider, for example, the case with $n = 2$ in which $\mathcal{P}_1$ is an ice cream and $\mathcal{P}_2$ is an umbrella (section 1.4.2). Pandora's preferences are then likely to change when the event E that the day is sunny is replaced by the event that it is raining. The decision problem then needs to be reformulated so that the consequences are identified with four of Pandora's possible states of mind. One such state would be how she feels when licking an ice cream on a sunny day. Another would be how she feels without an umbrella on a rainy day.

If we keep $E \neq \varnothing$ temporarily fixed, we can regard each V_E as a Von Neumann and Morgenstern utility function defined on the set C of consequences. Postulate 9 tells us that each such utility function represents the *same* preferences over the set of all lotteries with prizes in C. It follows from section 3.6 that these utility functions differ only in where they place the zero and the unit on their utility scales. In particular, for all nonempty events E and all prizes $\mathcal{P}$,

$$V_E(\mathcal{P}) = AV_B(\mathcal{P}) + B, \tag{7.5}$$

where the constants $A > 0$ and B don't depend on $\mathcal{P}$. In fact, $B = 0$, because we have normalized so that $V_E(\mathcal{L}) = V_B(\mathcal{L}) = 0$.

Recall from (7.4) that $u = V_B$ is Pandora's Von Neumann and Morgenstern utility function over the set C of consequences. We therefore have that

$$V_E(\mathcal{P}) = p(E)u(\mathcal{P}),$$

where the fact that the constant A depends on the event E has been recognized by writing $A = p(E)$. This fact takes us quite a long way, since we can now write the Von Neumann and Morgenstern utility for the gamble a of figure 7.1 as

$$U(a) = p(E)u(\mathcal{P}) + p(\sim E)u(\mathcal{L}), \tag{7.6}$$

because $u(\mathcal{L}) = V_B(\mathcal{L}) = 0$. Pandora therefore chooses among gambles as though maximizing the expected value of her Von Neumann and Morgenstern utility over the consequences in C relative to her subjective probabilities for the events in B. However, we have gotten ahead of ourselves by calling $p(E)$ a probability without asking whether the requirements of section 5.3.1 are satisfied. Our normalization of V_E ensures that $0 = p(\varnothing) \leqslant p(E) \leqslant p(B) = 1$ and so only the additivity of p is in doubt.

Finite additivity. Savage wouldn't have liked our building a theory of subjective probability around a theory of objective probability, but even de Finetti would have approved of our asking no more than finite additivity from subjective probabilities.

To this end, we now strengthen postulate 8 so as to incorporate an analogue of Savage's sure-thing principle.

Postulate 9*. If E isn't the empty event, the preferences revealed by Pandora's choices over lotteries of the form

$\mathcal{L}$		$\mathcal{P}_1$	$\mathcal{Q}$		$\mathcal{L}$		$\mathcal{P}_2$	$\mathcal{Q}$		$\mathcal{L}$	$\mathcal{P}_n$ $\mathcal{Q}$
$\sim(E \cup F)$		E	F		$\sim(E \cup F)$		E	F		$\sim(E \cup F)$	E F

$$q_1 \qquad\qquad q_2 \qquad \cdots \qquad q_n$$

remain unchanged if E is replaced everywhere by any other nonempty event that doesn't overlap with F.

Provided that we keep $\mathcal{Q}$ and F fixed, postulate 9* permits the analysis leading to (7.5) to go through with gambles like a of figure 7.1 replaced by gambles like b.

As before, the constants A and B are independent of $\mathcal{P}$, but now they depend on $\mathcal{Q}$ and F as well as E. Nor is it true that $B = 0$ in this new situation. Setting $\mathcal{P} = \mathcal{L}$ yields that $B = p(F)u(\mathcal{Q})$. Hence the Von Neumann and Morgenstern utility for gamble b in figure 7.1 is

$$U(b) = A(\mathcal{Q}, E, F)u(\mathcal{P}) + p(F)u(\mathcal{Q}). \qquad (7.7)$$

Similar reasoning also leads to the equation

$$U(b) = A(\mathcal{P}, F, E)u(\mathcal{Q}) + p(E)u(\mathcal{P}). \qquad (7.8)$$

It follows from (7.7) and (7.8) that

$$\frac{A(\mathcal{Q}, E, F) - p(E)}{u(\mathcal{Q})} = \frac{A(\mathcal{P}, F, E) - p(F)}{u(\mathcal{P})}.$$

Since the left-hand side is independent of $\mathcal{P}$ and the right-hand side is independent of $\mathcal{Q}$, both sides must be equal to a quantity $h(E, F)$ that depends on neither $\mathcal{P}$ nor $\mathcal{Q}$. Thus,

$$U(b) = p(E)u(\mathcal{P}) + p(F)u(\mathcal{Q}) + h(E, F)u(\mathcal{P})u(\mathcal{Q}). \qquad (7.9)$$

Take $\mathcal{P} = \mathcal{Q}$ in (7.9). Since $U(b)$ is then equal to $p(E \cup F)u(\mathcal{P})$, we obtain that

$$p(E \cup F)x = (p(E) + p(F))x + h(E, F)x^2, \qquad (7.10)$$

where $x = u(\mathcal{P})$. We have assumed that there are consequences $\mathcal{P}$ satisfying $\mathcal{L} \prec \mathcal{P} \prec \mathcal{W}$, so (7.10) holds for at least three values of x. But a quadratic equation can only have two roots unless its coefficients are all zero. Thus $h(E, F) = 0$ and

$$p(E \cup F) = p(E) + p(F).$$

It follows that p is finitely additive and so counts as a probability.

Maximizing expected utility. Because $h(E, F) = 0$ and $u(\mathcal{L}) = 0$, (7.9) can be written as

$$U(b) = p(E)u(\mathcal{P}) + p(F)u(\mathcal{Q}) + p(\sim(E \cup F)u(\mathcal{L}).$$

Thus, Pandora behaves as though maximizing expected utility when choosing among gambles of the form b. To ensure that she always behaves as though maximizing expected utility when choosing among gambles, we strengthen postulate 9* to postulate 9**.

*Postulate 9**. Without changing the conclusion of postulate 9*, we can replace the single event F that results in the single prize Q with a finite number of nonoverlapping events $F_1, F_2, \ldots, F_n$ that result in prizes $\mathcal{Q}_1, \mathcal{Q}_2, \ldots, \mathcal{Q}_n$.*

Our assumptions are adequate to ensure that Pandora also maximizes expected utility when choosing among compound gambles like (7.1), but this issue is put aside until we study Bayes' rule in section 7.4.

7.3 Dutch Books

Old wars between the English and the Dutch are commemorated by various pejorative sayings. A Dutch uncle is someone who keeps a mistress. Dutch courage is found in a bottle. A Dutch treat is when you have to pay for yourself. A Dutch book is a system of bets which guarantee that anyone who takes them all on will lose no matter what happens.

Dutch books arise in Bayesian decision theory because they provide an evolutionary defense of subjective probability analogous to the money-pump argument used to defend transitivity in section 1.6. It can be shown that if Pandora never falls prey to a Dutch book, then she behaves as Bayesian decision theory predicts.

Bookmaking. A Dutch book is the economic equivalent of the fabled Philosopher's Stone that transforms base metal into gold, but you don't need the crucibles and retorts with which an alchemist's laboratory is traditionally equipped to create a Bookmaker's Stone. All you need are one or more stubborn folk whose subjective probabilities are inconsistent.

Suppose, for example, that Alice is quite sure that the probability of *Punter's Folly* winning the Kentucky Derby is $\frac{3}{4}$. If she bets $3x$ at odds of $1:3$ against *Punter's Folly* winning, she therefore believes her expected dollar gain will be

$$-3x \times \tfrac{1}{4} + x \times \tfrac{3}{4} = 0.$$

So if the bookie offers her odds of $1:2$ against *Punter's Folly* winning, her expected dollar gain will be positive. For small enough values of x, she would therefore rather take the bet than not, provided that her Von Neumann and Morgenstern utility function for dollar gains is smooth.

Bob is quite sure that the probability of *Punter's Folly* winning the Kentucky Derby is only $\frac{1}{4}$. Since his Von Neumann and Morgenstern utility function for dollar gains is also smooth, the bookie offers him odds of $2:1$ against *Punter's Folly* winning. Like Alice, he will bet $\$3x$ at these odds for small enough values of x.

The bookie can now make a Dutch book against Alice and Bob. If each bets three cents, the bookie will always win three cents from one and pay two cents to the other. This is the secret of how bookies make money. Far from being wild gamblers as they like their customers to think, they bet only on sure things.

The bookie can exploit Alice and Bob because they have inconsistent beliefs. The Dutch book defense of Bayesian decision theory says that we can do the same to Pandora if some of her beliefs are inconsistent with each other.

7.3.1 Avoiding Dutch Books

Although it hasn't been given much emphasis, we have been assuming that Pandora's choice behavior reveals a full or complete set of preferences over all acts (section 1.5.2). In particular, she is able and willing to make a choice between any possible pair of acts. The Dutch book argument doesn't require the full force of this assumption, but what it does require remains very strong.

Pandora is invited to consider all possible bets about whether events in B will occur or not. For each possible bet, Pandora's feasible set is restricted to two alternatives. She can choose whether to be the bookie offering the bet or the punter to whom the bet is offered. That is to say, Pandora is never allowed the option of refusing a bet, but she gets to choose which side of the bet to back.

A difficulty in the original Dutch book formulations of Ramsey (1931) and de Finetti (1937) is that their bets were denominated in dollars. One can eliminate this problem by denominating bets in notional poker chips, each of which corresponds to one util on Pandora's Von Neumann and Morgenstern utility scale, but this seems to me to be a bit pedantic.

To illustrate the way the Dutch book argument goes, consider bets in which odds of $g:1$ are offered against the event E occurring. For each such bet, Pandora chooses whether to be the punter or the bookie. If she chooses to be the bookie when $g = a$ and the punter when $g = b$, we must have $a \leqslant b$ or Pandora will be caught in a Dutch book.[3] So we can find odds of $c : 1$ such that Pandora chooses to be the bookie when $g < c$ and the punter when $g > c$. She is then acting as though her subjective probability for the event E is $p(E) = 1/(c + 1)$.

When the state E arises in other gambles, Pandora must continue to behave as though its probability were $p(E)$, otherwise a Dutch bookie will exploit the fact that she sometimes assigns one probability to E and sometimes another. Nor must Pandora neglect to manipulate her subjective probabilities according to the standard laws of probability lest further Dutch books be made against her.

If Pandora always takes one side or the other of every bet, she must therefore honor Bayesian decision theory or be vulnerable to a Dutch book.

Completeness and consistency. It isn't very realistic to insist that Pandora must always bet one way or another no matter what level of confidence she may have in her beliefs. Such a lack of realism represents a major weakness in the Dutch book argument, particularly when the idea is applied to logical probabilities (Howson 2000). It also points up a lesson that will be important in later chapters.

There is always a tension between completeness and consistency (section 8.4). The difficulties that arise in decision theory because of this tension are only a pale shadow of the difficulties that Gödel traced to this source in the foundations of mathematics, but we have to deal with them nevertheless.

We insist on completeness in the Dutch book argument by forcing Pandora to commit herself to making a choice from any feasible set we care to name. We insist on consistency by demanding that she be invulnerable to any Dutch book we care to construct. Both demands together force Pandora into a Bayesian straitjacket. But do we really want to insist that she always has a subjective probability to everything—even for events about which she is entirely ignorant?

If we abandon completeness by allowing Pandora to refuse to bet in situations in which she has little confidence in her ability to guess successfully, we are led to the quasi-Bayesian theory of Giron and Rios (1980) with upper and lower subjective probabilities (section 5.6.3). If we insist on completeness but accept that Pandora may be averse to uncertainty, then we have to give up the strong consistency requirements built into

---

[3] If she chooses to be the bookie for all values of g, then she acts as though E is impossible. If she chooses to be the punter for all values of g, then she acts as though E is certain.

Bayesian decision theory. In later chapters, I therefore (very reluctantly) surrender the requirement that rationality always requires separating your preferences from your beliefs (section 9.2).

7.4 Bayesian Updating

The ghost of the Reverend Thomas Bayes must be in a constant state of astonishment that our culture has embraced a philosophical doctrine called Bayesianism that treats the trivial manipulation of conditional probabilities that he discovered sometime before 1764 as the solution to the problem of scientific induction.

How is it possible that a rule justified by appealing to the properties of frequencies could be applied so thoughtlessly to updating degrees of belief (section 6.3.3)? If we told him that Pandora's subjective probabilities must obey Bayes' rule (and all the other laws of probability) if she is to escape falling prey to a Dutch book, he would doubtless agree, but ask why anyone should think that Pandora's betting behavior should be thought relevant to assessing scientific propositions. The rest of this chapter is an account of how Savage would have answered this question in the context of a small world.

7.4.1 Justifying Bayes' Rule

Postulate 7 says that Pandora doesn't care about how gambles are put together. Only the bottom line matters to her.[4] It follows that

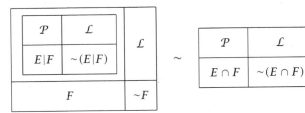

where $E \mid F$ is defined as the event that F first occurs and then E. Savage's theory of subjective probability then forces us to accept Kolmogorov's definition of a conditional probability:

$$p(E \mid F)p(F) = p(E \cap F).$$

Bayes' rule then follows in the usual way (section 5.5).

---

[4] We made a similar assumption in postulates 3 and 4 of chapter 3, but there it was necessary to insist that all the lotteries involved were independent so that we could write down a simple formula for the probability of each final outcome.

7.4.2 Priors and Posteriors

Making sensible inferences from new information is one of the most important aspects of rational behavior. Bayesian updating is how such inferences are made in Bayesian decision theory.

The language of *prior* and *posterior* probabilities is often used when discussing such inferences. Your prior probabilities quantify your beliefs before something happens. Your posterior probabilities quantify your beliefs after it has happened.

Tossing coins. The probability that a weighted coin falls *heads* is p. Pandora's prior probabilities over the possible values of p are $\text{prob}(p = \frac{1}{3}) = 1 - q$ and $\text{prob}(p = \frac{2}{3}) = q$. (Values of p other than $\frac{1}{3}$ and $\frac{2}{3}$ are assumed impossible.) What are her posterior probabilities after observing the event E in which *heads* appears m times, and tails n times in $N = m + n$ tosses?

When applying Bayes' rule to such a problem, it is usual to simplify things by postponing the calculation of $C = 1/\text{prob}(E)$. We write down the conditional probability given E of all the events into which we have partitioned the possibilities, and then compute C using the fact that these probabilities must sum to one. In Pandora's case:[5]

$$\text{prob}(p = \tfrac{2}{3} \mid E) = C \, \text{prob}(E \mid p = \tfrac{2}{3}) \, \text{prob}(p = \tfrac{2}{3}) = \frac{2^m q}{2^m q + 2^n (1 - q)},$$

$$\text{prob}(p = \tfrac{1}{3} \mid E) = C \, \text{prob}(E \mid p = \tfrac{1}{3}) \, \text{prob}(p = \tfrac{1}{3}) = \frac{2^n (1 - q)}{2^m q + 2^n (1 - q)}.$$

What if $m \approx \frac{2}{3} N$ and $n \approx \frac{1}{3} N$, so that the frequency of heads is around $\frac{2}{3}$? If N is large, we would regard this as evidence that the objective probability of the coin falling heads is about $\frac{2}{3}$. Pandora's posterior probability that $p = \frac{2}{3}$ is correspondingly close to one, because

$$\text{prob}(p = \tfrac{2}{3} \mid E) \approx \frac{q}{q + (1 - q) 2^{-N/3}} \to 1 \quad \text{as } N \to \infty.$$

This example illustrates a relation between subjective and objective probabilities to which Bayesians understandably attach much importance. Unless Pandora's prior assigns zero probability to the true value of a probability p, her posterior probability for p will be approximately one with high probability after observing enough independent trials.

The argument by design. Bayesianism holds that Bayes' rule can be applied not just to coin tossing and the like, but to anything whatever—including the argument by design that some theologians defend as a valid demonstration of the existence of God.

---

[5] The binomial distribution tells us that the probability of exactly m heads in $m + n$ tosses when *heads* falls with probability p is $(m + n)! p^m (1 - p)^n / m! n!$.

Let F be the event that something appears to have been organized. Let G be the event that there is an organizer. Everybody agrees that $\mathrm{prob}(F\mid G) > \mathrm{prob}(F\mid \sim G)$, but the argument by design needs to deduce that $\mathrm{prob}(G\mid F) > \mathrm{prob}(\sim G\mid F)$ if God's existence is to be more likely than not. Applying Bayes' rule, we find that

$$\mathrm{prob}(G\mid F) = \frac{\mathrm{prob}(F\mid G)\,\mathrm{prob}(G)}{\mathrm{prob}(F\mid G)\,\mathrm{prob}(G) + \mathrm{prob}(F\mid \sim G)\,\mathrm{prob}(\sim G)}.$$

If $\mathrm{prob}(G) \geqslant \mathrm{prob}(\sim G)$, we can deduce the required conclusion that $\mathrm{prob}(G\mid F) > \mathrm{prob}(\sim G\mid F)$, but we are otherwise left in doubt.

Bayesianism therefore has an explanation of why religious folk are more ready to accept the argument by design than skeptics!

7.4.3   Where Do Priors Come From?

If Bayesianism were right in claiming that Bayes' rule solves the problem of scientific induction, then updating your beliefs when you get new information would simply be a matter of carrying out some knee-jerk arithmetic. But what of the prior probabilities with which you begin? Where do they come from? Why should you stick with the prior you started with after being surprised by some data you didn't anticipate, rather than starting up anew with a fresh prior?

The Harsanyi doctrine.   The gift of rationality is sometimes said to include a metaphysical hotline to the choice of the correct prior.

We met John Harsanyi's even stronger form of this claim in section 4.4.1. He advocated using a mind experiment to determine the one-and-only rational prior. If Pandora imagines that a veil of ignorance conceals all the information she has ever received, she will supposedly select the same prior as all other ideally rational folk in the same state of sublime ignorance. But when I try this trick, no ideas about a suitable prior come to me at all.

The principle of insufficient reason.   Bayesian statisticians use their experience of what has worked out well in the past when choosing a prior. Bayesian physicists prefer whatever prior maximizes entropy (Jaynes and Bretthorst 2003). Otherwise, an appeal is usually made to the principle of insufficient reason (also called the principle of of indifference). This principle—which is variously attributed to Laplace or Jacob Bernoulli—says that Pandora should assign the same probability to two events if she has no reason to believe that one is more likely to occur than the other. But the principle is painfully ambiguous.

For example, what prior should we assign to Pandora when she knows nothing at all about the three horses running in a race? Does the principle of insufficient reason tell us to give each horse a prior probability of $\frac{1}{3}$? Or should we give a prior probability of $\frac{1}{2}$ to *Punter's Folly*, because Pandora has no reason to think it more likely that *Punter's Folly* will win than lose?

The geometric examples of section 5.5.3 provide some less frivolous examples (Keynes 1921), but my favorite is the wine–water paradox (Gillies 2000, p. 38).

The wine–water paradox. Some wine and water have been mixed so that there is no more than three times one of the liquids than the other. What is the probability that there is no more than twice as much wine as water?

We are looking for the probability that

$$r = \frac{\text{wine}}{\text{water}} \leqslant 2.$$

The principle of insufficient reason would seem to imply that our prior for r should be the uniform probability distribution on the interval $[\frac{1}{3}, 3]$, and so

$$\text{prob}(r \leqslant 2) = \frac{2 - \frac{1}{3}}{3 - \frac{1}{3}} = \frac{5}{8}.$$

But we could equally well have asked for the probability that

$$s = \frac{\text{water}}{\text{wine}} \geqslant \frac{1}{2}.$$

The principle of insufficient reason would again seem to imply that our prior for s should be the uniform probability distribution on the interval $[\frac{1}{3}, 3]$, and so

$$\text{prob}(s \geqslant \tfrac{1}{2}) = \frac{3 - \frac{1}{2}}{3 - \frac{1}{3}} = \frac{15}{16}.$$

Applying the principle of insufficient reason to two different ways of expressing the same problem therefore leads to two different answers.

7.5 Constructing Priors

Savage's formal theory of subjective probability is silent on the subject of priors. It says that, if Pandora chooses consistently, then she chooses as though she had quantified her beliefs using subjective probabilities. But where did her beliefs come from in the first place? What is the origin of the prior probabilities that she updates using Bayes' rule as she gains experience?

To seek to answer such questions is to move out of the arena of subjective probability into the arena of logical or epistemic probability. Bayesianism treats this step as unproblematic, but Savage was more cautious.

Gut feelings? It is common to register our lack of understanding of how Pandora converts her general experience of the world into subjective beliefs by saying that the latter reflect her "gut feelings." But she would be irrational to treat the rumblings of her innards as an infallible oracle. Our gut feelings are usually confused and inconsistent. When they uncover such shortcomings in their beliefs, intelligent people modify the views about which they are less confident in an attempt to bring them into line with those about which they are more confident.

Savage (1951, pp. 100–4) thought that his theory would be a useful tool for this purpose. His response to Allais mentioned in section 3.5 illustrates his attitude. When Allais pointed out an inconsistency in his choices, Savage recognized that his gut had acted irrationally, and modified his behavior accordingly. Luce and Raiffa's (1957, p. 302) *Games and Decisions* summarizes his approach as follows:

> Once confronted with inconsistencies, one should, so the argument goes, modify one's initial decisions so as to be consistent. Let us assume that this jockeying—making snap judgments, checking up on their consistency, modifying them, again checking on consistency, etc.—leads ultimately to a bona fide, prior distribution.

I agree with Savage that, without going through such a process of reflective introspection, there is no particular virtue in being consistent at all. But when the world in which we are making decisions is large and complex, there is no way that such a process could be carried through successfully. So Savage restricted the sensible application of his theory to small worlds.

7.5.1   Achieving Consistency in a Small World

How would Savage have arrived at a prior probability measure via the adjustment process that he envisaged being used to achieve consistency in a small world?

Figure 7.2 illustrates the advice I think Savage would have given to Pandora. He would explain that it is sensible for Pandora to consult her gut feelings when she has more evidence rather than less. For each possible future course of events, she should therefore ask herself, "What subjective probabilities would my gut come up with *after* experiencing these events?" In the likely event that these posterior probabilities turn out to

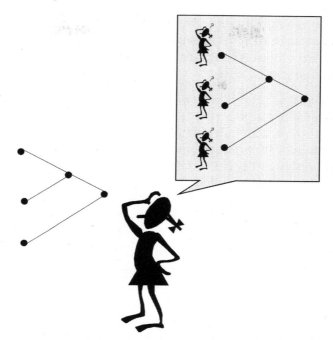

Figure 7.2. Constructing priors. Pandora faces an information tree that admits three possible future histories. As she progresses through the tree, she will update her prior probabilities using Bayes' rule. But how does she choose her prior? She would do better to consult her gut feelings with more information rather than less, so she imagines what subjective probabilities would seem reasonable after experiencing each possible future history. The resulting unmassaged posteriors are unlikely to be consistent. She therefore uses her confidence in the various judgments they incorporate to massage the posteriors until they become consistent. But consistent posteriors can be deduced by Bayes' rule from a prior. Instead of deducing her posterior probabilities from prior probabilities chosen on a priori grounds, Pandora therefore constructs her prior by massaging the unmassaged posterior probabilities with which her analysis begins.

be inconsistent with each other, she should then take account of the confidence she has in her initial snap judgments to massage her posterior probabilities until they become consistent. After the massaging is over, Pandora would then be invulnerable to surprise, because she would have *already* taken account of the impact that any future information might have on the internal model that she uses in determining her beliefs.

The end product of Pandora's massaging process is a bunch of consistent posterior probabilities. In Savage's theory of subjective probability, this means that they can all be deduced from the same prior probabilities. Pandora can therefore summarize her massaged posterior probabilities by working out the prior probabilities from which they can be deduced. In this story, Bayes' rule is therefore reduced to nothing more than a bookkeeping tool that saves Pandora from having to remember all her massaged posterior probabilities.

Reversing Bayesianism. It is ironic that the avowed prophet of Bayesianism should advocate a procedure that reverses what Bayesianism recommends. Instead of deducing our posteriors from a metaphysical prior distilled somehow from the air, we are advised to deduce our prior from a massaged set of posteriors. I think this is why naive Bayesians neglect Savage's warning that his theory of subjective probability only makes sense in a small world. If you don't think it necessary to bring some kind of massaging process to a successful conclusion in coming up with a prior, why should it matter whether the world is large or small?

7.5.2   Rational Learning?

Rational learning is sometimes said to consist of no more than applying Bayes' rule, but this isn't true in Savage's massaging story. Pandora must certainly take note of what happens as she applies Bayes' rule on receiving new pieces of information, but such bookkeeping isn't genuine learning. Otherwise it wouldn't seem absurd to say that a camera is learning the images it records.

Pandora does her learning in advance of receiving any information when she takes into account the effect that possible future surprises may have on the unformalized internal model that she uses to construct her beliefs. She learns as she refines her beliefs during the massaging process. Bayesian updating only takes place after all genuine learning is over.

The idea that Bayesian updating in a small world involves no genuine learning at all commonly provokes incredulity. Am I saying that we can only learn when deliberating about the future, and never directly from experience? The brief answer is *no*, but I have learned directly from experience that a longer answer is necessary.

Learning from experience?   The fact that real people actually learn from experience isn't relevant to whether Bayesian updating in a small world should count as rational learning. The worlds about which real people learn are almost always large. Even when they are confronted with a small world in laboratories, ordinary people almost never use Bayesian updating.

Without training, even clever people turn out to be remarkably inept in dealing with simple statistical problems involving conditional probabilities (section 5.5). In my own game theory experiments, no subject has ever given a Bayesian answer to the question "Why did you do what you did?" when surveyed after the experiment—even though the majority were drawn from populations of students who had studied Bayesian statistics.

Professional statisticians are an exception to this generalization. Bayesian statisticians don't ask themselves why a knee-jerk adherence to consistency requirements is appropriate, but simply update from a prior distribution chosen on the basis of what their past experience has shown to work out well in analogous situations. I don't argue that proceeding in this way is unreasonable. Nor do I argue that a Bayesian statistician who updates from a prior distribution chosen on a priori grounds isn't learning. But when Bayesian statisticians learn in this way, they aren't working in a small world, and so they can't derive any authority for their procedures from Savage's theory. In particular, they have no grounds for claiming that their methodology is superior to classical statistics.

The problem of how best to learn in a large world is unsolved. As long as it remains unsolved, any learning procedures that we employ in the context of a large world will necessarily remain arbitrary to some extent. My guess is that the problem of scientific induction will always remain unsolved, because it is one of those problems that has no definitive solution.

Learning arithmetic? We aren't through with the question of whether Bayesian updating in a small world can properly count as learning. So far, it has only been argued that the fact that ordinary people and Bayesian statisticians learn from experience is irrelevant. To take the argument forward, I want to draw an analogy between how a Bayesian learns while using the massaging methodology I have attributed to Savage, and how a child learns arithmetic.

When Alice learns arithmetic at school, her teacher doesn't know what computations life will call upon her to make. Among other things, he therefore teaches her an algorithm for adding numbers. This algorithm requires that Alice memorize some addition tables. In particular, she must memorize the answer to $2 + 3 = ?$. If the teacher is good at his job, he will explain why $2 + 3 = 5$. If Alice is an apt pupil, she will understand his explanation. One may then reasonably say that Alice has learned that, should she ever need to compute $2 + 3$, then the answer will be 5.

Now imagine Alice in her maturity trying to complete an income tax form. In filling the form, she finds herself faced with the problem of computing $2 + 3$, and so she writes down the answer 5. Did she just learn that the answer to this problem is 5? Obviously not. She learned this in school. All that one can reasonably say that she "learned" in filling the form is that filling the form requires computing $2 + 3$.

Bob was such a poor student that he never learned the addition tables. While filling the tax form, he might perhaps use his fingers to reckon with, and thereby discover that $2 + 3 = 5$. He would then undoubtedly have learned something. But he wouldn't be operating in a small world, within which all potential surprises have been predicted and evaluated in advance of their occurrence.

Ramsey on learning.   Colin Howson (2000, p. 145) identifies what he regards as a "rare slip" in Frank Ramsey's (1931) pioneering discussion of subjective probability. In one place, Ramsey says that it would be inconsistent for him to alter the value of $\text{prob}(E \mid F)$ after observing F. Elsewhere, he says that learning F might profoundly alter his whole system of beliefs for psychological reasons. Howson, who is engaged in a brave attempt to interpret probability as a logical degree of belief, believes that Ramsey should have accepted that learning F might make it necessary to alter one's whole system of beliefs for logical reasons.

I don't know what would be true of an adequate theory of logical probability, but I think Ramsey was right on target in the case of subjective probability. After carrying through Savage's massaging process, Pandora will already have anticipated what her reaction would be if she were to learn F. She will therefore have no reason to alter $\text{prob}(E \mid F)$ should F occur.

On the other hand, if she hasn't carried through some analogue of the massaging process, there is no particular reason why she should be consistent at all. It may be that the occurrence of F inspires her with a whole new worldview. She may previously have been using Bayes' rule to update her probabilities from a prior derived from her old worldview, but she now believes her old worldview to be wrong. She will then ask what virtue there is in being consistent if the result is that she is consistently wrong. If she remains a victim of Bayesianism, why shouldn't she go back to the beginning and start updating all over again from whatever new prior is suggested by her new worldview? The value of $\text{prob}(E \mid F)$ at which she then arrives might be anything at all.

7.6   Bayesian Reasoning in Games

The models constructed by game theorists are small worlds almost by definition. So we can use Bayesian decision theory without fear of being haunted by Savage's ghost telling us that it is ridiculous to use his theory in a large world. Bayesian decision theory only ceases to be appropriate when attempts are made to include the minds of the players in the model to be considered.

7.6.1 Subjective Equilibria

Very naive Bayesians don't see any need to worry about equilibria in games. When playing Matching Pennies, so their story goes, Alice's gut feelings tell her what subjective probabilities to assign to Bob's choosing *heads* or *tails*. She then makes her own choice of *heads* or *tails* to maximize her expected utility. Bob proceeds in the same way. The result may not be an equilibrium, but so what?

But it isn't so easy to escape the problems raised by sentences that begin: "Alice thinks that Bob thinks that Alice thinks ...." In forming her own subjective beliefs about Bob, Alice will simultaneously be trying to predict how Bob will form his subjective beliefs about her. While using something like the massaging process of section 7.5.1, she will then not only have to massage her own probabilities until consistency is achieved, she will also have to simulate Bob's similar massaging efforts. The end product will include not only Alice's subjective probabilities for Bob's choice of strategy, but her prediction of his subjective probabilities for her choice of strategy. The two sets of subjective probabilities must be consistent with the fact that both players will optimize on the basis of their subjective beliefs. If so, we are looking at a Nash equilibrium. If not, a Dutch book can be made against Alice (Nau and McCardle 1990).

When a Nash equilibrium is interpreted in this way—so that beliefs rather than strategies are treated as primary—it is called a *subjective equilibrium*. Without this notion, it would be hard to find a rational interpretation of mixed Nash equilibria.

For example, the Nash equilibrium of Matching Pennies requires that Alice and Bob both play *heads* and *tails* with probability $\frac{1}{2}$ (section 6.2). But, given the other's play, they are both indifferent between all their strategies, so why choose the particular strategy required by the Nash equilibrium? However, if we reinterpret the Nash equilibrium as an equilibrium in beliefs, the problem goes away; the only stable belief that each player can hold is that each is equally likely to play *heads* or *tails*.

7.6.2 Common Priors

Games often involve chance moves about which the players are assumed to have subjective beliefs. The argument justifying subjective equilibria still applies, but if Alice is to avoid a Dutch book based on her predictions of everybody's beliefs, her massaging efforts must generate a common prior from which everybody's posterior beliefs can be deduced by conditioning on their information.

Game theorists often assume that all players will necessarily be led to attribute the *same* common prior to each other. The Harsanyi doctrine is designed to allow this conclusion (section 7.4.3), but there is no reason why every player should be led to the same common prior if we assume that they achieve consistency using something like Savage's massaging process.

In complicated games, one can expect the massaging process to converge on the same common prior for all players only if their gut feelings are similar. But we can only expect the players to have similar gut feelings if they all share a common culture, and so have a similar history of experience. Or to say the same thing another way, only when the players of a game are members of a reasonably close-knit community can they be expected to avoid leaving themselves open to a Dutch book being made against their group as a whole.

Epistemology

8.1 Knowledge

Philosophers traditionally treat knowledge as justified true belief, and then argue about what their definition means. This chapter contributes little to this debate, because it defends an entirely different way of thinking about knowledge. However, before describing the approach to knowledge that I think most useful in decision theory, it is necessary to review Bayesian epistemology—the study of how knowledge is treated in Bayesian decision theory.

8.2 Bayesian Epistemology

When Pandora uses Bayes' rule to update her prior probability $\text{prob}(E)$ of an event E to a posterior probability $\text{prob}(E \mid F)$ on learning that the event F has occurred, everybody understands that we are taking for granted that she has received a signal that results in her knowing that F has occurred. For example, after Pandora sees a black raven in Hempel's paradox, she deduces that the event F of figure 5.1 has occurred. However, when deciding what Pandora may reasonably be assumed to know after receiving a signal, Bayesians assume more than they often realize about how rational agents supposedly organize their knowledge.

This section therefore reviews what Bayesian decision theory implicitly assumes about what Pandora can or can't know if she is to be regarded as consistent. The simple mathematics behind the discussion are omitted since they can be found in chapter 12 of my book *Playing for Real* (Binmore 2007b).

Knowledge operators. We specify what Pandora knows with a knowledge operator $\mathcal{K}$. The event $\mathcal{K}E$ is the set of states of the world in which Pandora knows that the event E has occurred. So $\mathcal{K}E$ is the event that Pandora knows E. Not knowing that something didn't happen is equivalent to thinking it possible that it did happen. So the possibility operator is defined by $\mathcal{P}E = \sim\mathcal{K}\sim E$ (where $\sim F$ is the complement of F).

(K0)	$\mathcal{K}B = B$	(P0)	$\mathcal{P}\varnothing = \varnothing$
(K1)	$\mathcal{K}(E \cap F) = \mathcal{K}E \cap \mathcal{K}F$	(P1)	$\mathcal{P}(E \cup F) = \mathcal{P}E \cup \mathcal{P}F$
(K2)	$\mathcal{K}E \subseteq E$	(P2)	$\mathcal{P}E \supseteq E$
(K3)	$\mathcal{K}E \subseteq \mathcal{K}^2 E$	(P3)	$\mathcal{P}E \supseteq \mathcal{P}^2 E$
(K4)	$\mathcal{P}E \subseteq \mathcal{K}\mathcal{P}E$	(P4)	$\mathcal{K}E \supseteq \mathcal{P}\mathcal{K}E$

Figure 8.1. The knowledge operator. The five properties that characterize the knowledge operator are jointly equivalent to the five properties that characterize the possibility operator. Properties (K0) and (K3) are strictly redundant, because they can be deduced from the other knowledge properties. One can also replace $\subseteq$ in (K3) and (K4) by =. If B is infinite, then (K1) needs to be modified to allow $\mathcal{K}$ to commute with infinite intersections.

The properties of knowledge and possibility operators usually taken for granted in Bayesian decision theory are listed in figure 8.1 for a finite state space B.[1]

Consistency and completeness. Property (K0) looks innocent, but it embodies a completeness assumption that Pandora may not always be willing to make. It says that she always knows the universe of discourse she is working in. But do we want to say that our current language will always be adequate to describe all future scientific discoveries? Are we comfortable in claiming that we have already conceived of all possibilities that we may think of in the future?

Property (K2) is a consistency assumption. It says that Pandora is infallible—she never knows something that isn't true. But do we ever know what is really true? Even if we somehow hit upon the truth, how could we be sure we were right?

Both the completeness assumption and the consistency assumption are therefore open to philosophical doubt. But we shall find that matters are worse when we seek to apply them simultaneously in the kind of large world to which Bayesian decision theory doesn't apply (section 8.4). The tension between completeness and consistency mentioned in section 7.3.1 then becomes so strong that one or the other has to be abandoned.

Knowing that you know. Thomas Hobbes must have been desperate to rescue a losing argument in 1654 when he asked René Descartes whether we really know that we know that we know that we know something. If Descartes had been a Bayesian, he would have replied that it is

---

[1] They are equivalent to the properties attributed to the operators □ and ◊ by the modal logic S-5. It is usual in epistemic logic to write the requirements for S-5 differently. They are standardly written in terms of propositions (rather than events) and include the assumption that $\mathcal{K}(\phi \rightarrow \psi) \rightarrow (\mathcal{K}\phi \rightarrow \mathcal{K}\psi)$ (Hintikka 1962).

only necessary to iterate (K3) many times to ensure that someone who knows something knows that they know to any depth of knowing that we choose.

Not knowing that you don't know. Donald Rumsfeld will probably be forgotten as one of the authors of the Iraq War by the time this book is published, but he deserves to go down in history for systematically considering all the ways of juxtaposing the operators $\mathcal{K}$ and $\sim\!\mathcal{K}$ at a news conference. When he got to $\sim\!\mathcal{K}\sim\!\mathcal{K}$, he revealed that he was no Bayesian by failing to deduce from (K4) that $\sim\!\mathcal{K}\sim\!\mathcal{K}E = \mathcal{K}E$.

The claim that we necessarily know any facts that we don't know that we don't know is hard to swallow. But the logic that supports $\sim\!\mathcal{K}\sim\!\mathcal{K}E = \mathcal{K}E$ in a small world is inescapable. Pandora doesn't know that she doesn't know that she has been dealt the Queen of Hearts. But she would know she hadn't been dealt the Queen of Hearts if she had been dealt some other card. She therefore knows that she wasn't dealt some other card. So she knows she was dealt the Queen of Hearts.

All things considered. The preceding discussion makes it clear that Bayesian epistemology only makes good sense when Pandora has carefully thought through all the logical implications of all the data at her disposal. Philosophers sometimes express this assumption of omniscience by saying Pandora's decisions are only taken after "all things have been considered."

However, Donald Rumsfeld was right to doubt that all things can be considered in a large world. In a large world, some bridges can only be crossed when you come to them. Only in a small world, in which you can always look before you leap, is it possible to consider everything that might be relevant to the decisions you take.

8.3 Information Sets

How does Pandora come to know things? In Bayesian epistemology, Pandora can only know something if it is implied by a truism, which is an event that can't occur without Pandora knowing it has occurred.

The idea is more powerful in the case of *common* knowledge, which is the knowledge that groups of people hold in common. The common knowledge operator satisfies all the properties listed in figure 8.1, and so nothing can be common knowledge unless it is implied by a public event—which is an event that can't occur without it becoming common knowledge that it has occurred. Public events are therefore rare.

They happen only when we all observe each other observing the same occurrence.

Possibility sets. The smallest truism $P(s)$ containing a state s of the world is called a possibility set. The event $P(s)$ consists of all the states t that Pandora thinks are possible when the actual state of the world is s. So if s occurs, the events that Pandora knows have occurred consist of $P(s)$ and all its supersets.

The assumptions of Bayesian epistemology can be neatly expressed in terms of possibility sets. It is in this form that they are commonly learned by Bayesians—who are sometimes a little surprised to discover that they are equivalent to (K0)-(K4). The equivalence is established by defining $\mathcal{P}$ and P in terms of each other using the formula

$$t \in P(s) \quad \Leftrightarrow \quad s \in \mathcal{P}\{t\}.$$

The formulation of Bayesian epistemology in terms of possibility sets requires only two assumptions:

(Q1) *Pandora never thinks the true state is impossible*;

(Q2) *Pandora's possibility sets partition the state space B.*

The requirement (Q1) is the infallibility or consistency requirement that we met as (K2) in section 8.2. It says that s always lies in $P(s)$. To partition B is to break it down into a collection of subsets so that each state in B belongs to one and only one subset in the collection (section 5.2.1). Given (Q1), the requirement (Q2) therefore reduces to demanding that $P(s)$ and $P(t)$ either have no states in common or else are different names for the same set.

One can think of Pandora's possibility sets as partitioning her universe of discourse into units of knowledge. When the true state is determined, Pandora necessarily learns that one and only one of these units of knowledge has occurred. Everything else she knows can then be deduced from this fact.

8.3.1   Applications to Game Theory

A case can be made for crediting John Von Neumann with the discovery of Bayesian epistemology. The information sets he introduced into game theory are versions of possibility sets modified to allow updating over time to be carried out conveniently. When first reading Von Neumann and Morgenstern's (1944) *Theory of Games and Economic Behavior*, I recall being astonished that so many problems with time and information could be handled with such a simple and flexible formalism.

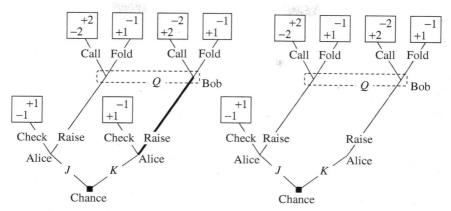

Figure 8.2. Mini-Poker. Alice is dealt a *Jack* or a *King* with equal probability. She sees her card but Bob doesn't. Everybody knows that Bob has been dealt a *Queen*. The left-hand game tree shows the betting rules. Bob only gets to act if Alice raises. He then knows that Alice has raised, but not whether she holds a *Jack* or a *King*. This information is represented by the information set labeled Q. Bob knows he is at one of the two nodes in Q, but not whether he is at the left node or the right node. It is always optimal for Alice to raise when dealt a *King*, and so this choice has been thickened in the left-hand game tree to indicate that Bob *believes* that she will raise with probability one when holding a *King*. If Bob *knew* that she never checks when holding the *King*, this fact would be indicated as in the right-hand game tree by omitting this possibility from the model.

Mini-Poker. Translating Von Neumann's formalism into the official language of epistemic logic is nearly always a very bad idea when trying to solve a game, since epistemic logic isn't designed for practical problem-solving. However, this section goes some way down this road for the game of Mini-Poker, whose rules are illustrated in figure 8.2.[2]

All poker games begin with the players contributing a fixed sum of money to the pot. In Mini-Poker, Alice and Bob each contribute one dollar. Bob is then dealt a *Queen* face up. Alice is then dealt either a *Jack* or a *King* face down. Alice may then check or raise one dollar. If she checks, the game is over. Both players show their cards, and whoever has the higher card takes the pot. If Alice raises, Bob may call or fold. If he folds, Alice takes the pot regardless of who has the better card. If Bob calls, he must join Alice in putting an extra dollar in the pot. Both players then show their cards, and whoever has the higher card takes the pot.

---

[2] Mini-Poker is a simplification of a version of poker analyzed in Binmore (2007b, chapter 15) in which Alice and Bob are each dealt one card from a shuffled deck containing only the King, Queen, and Jack of Hearts. This game is itself a simplification of Von Neumann's (1944) second model of poker in which Alice and Bob are each dealt a real number between 0 and 1.

The most significant feature of the rules of the game illustrated in figure 8.2 is the information set labeled Q. If this information set is reached, then Bob knows he is at one of the two decision nodes in Q, but not whether he is at the left node or the right node. The fact that he knows Q has been reached corresponds to his knowing that Alice has raised the pot. The fact that he doesn't know which node in Q has been reached corresponds to his not knowing whether Alice has been dealt a *Jack* or a *King*.

How should Alice and Bob play if the *Jack* or the *King* is dealt with equal probability and both players are risk neutral?[3] Alice should always raise when holding the *King*. When holding the *Jack* she should raise with probability $\frac{1}{3}$. (Some bluffing is necessary, because Bob would always fold if Alice never raised when holding a *Jack*.) Bob should keep Alice honest by calling with probability $\frac{2}{3}$.

It would be too painful to deduce this result using the formal language of possibility sets, but it isn't so hard to check that Bob is indifferent between calling and folding if Alice plays according to the game-theoretic solution. (If he weren't indifferent, the solution would be wrong, since it can never be optimal for Bob to sometimes play one strategy and sometimes another if he believes one of them is strictly better than the other.)

Bob's possibility sets.   In principle, the states of the world in a game are all possible paths through the tree. In Mini-Poker, there are six such plays of the game, which can be labeled Jc, Jrc, Jrf, Kc, Krc, and Krf. For example, Krf is the play in which Alice is dealt a *King* and raises, whereupon Bob folds.

As the game proceeds, Alice and Bob update their knowledge partitions as they learn things about what has happened in the game so far. For example, figure 8.3 shows Bob's possibility sets after Alice's move, both before and after he makes his own move. (If he folds, Alice won't show him what she was dealt, because only novices reveal whether or not they were bluffing in such situations.)

Bayesian updating.   Bob believes that Alice is dealt a *Jack* or a *King* with probability $\frac{1}{2}$. He also believes that Alice will play according to the recommendations of game theory and so raise with probability $\frac{1}{3}$. His prior probabilities for the events Jc, Jr, Kc, Kr in figure 8.3 are therefore

$$\text{prob}(Jc) = \tfrac{1}{2} \times \tfrac{2}{3} = \tfrac{1}{3}, \qquad \text{prob}(Jr) = \tfrac{1}{2} \times \tfrac{1}{3} = \tfrac{1}{6},$$
$$\text{prob}(Kc) = \tfrac{1}{2} \times 0 = 0, \qquad \text{prob}(Kr) = \tfrac{1}{2} \times 1 = \tfrac{1}{2}.$$

---

[3] We solve such problems by computing Nash equilibria. Since Mini-Poker is a two-person zero-sum game, this is the same as finding the players' maximin strategies.

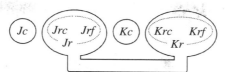

Bob's possibility sets after Alice's
move and before his own move

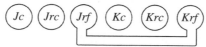

Bob's possibility sets after his move

Figure 8.3. Possibility sets for Bob in Mini-Poker. When Bob observes Alice raise the pot, he knows that the outcome of the game will be one of the four plays Jrc, Jrf, Krc, Krf. He will himself determine whether he calls or folds, and so he is only interested in the conditional probability of the events that Alice was dealt a *Jack* or a *King* given that he has observed her raise the pot. These are the events $Jr = \{Jrc, Jrf\}$ and $Kr = \{Krc, Krf\}$.

After Bob learns that Alice has raised, his possibility set becomes $F = \{Jr, Kr\}$. His probabilities for the events Jr and Kr have to be updated to take account of this information. Applying Kolmogorov's definition of a conditional probability (5.4), we find that Bob's posterior probabilities are

$$\text{prob}(Jr \,|\, F) = \frac{\frac{1}{6}}{\frac{1}{6} + \frac{1}{2}} = \frac{1}{4}, \qquad \text{prob}(Kr \,|\, F) = \frac{\frac{1}{2}}{\frac{1}{6} + \frac{1}{2}} = \frac{3}{4}.$$

Bob's expected payoff if he calls a raise by Alice is therefore $\frac{1}{4} \times 2 + \frac{3}{4} \times (-2) = -1$, which is the same payoff he gets from folding. He is therefore indifferent between the two actions.

Perfect recall. Von Neumann overlooked the need that sometimes arises to insist that rational players never forget any information. However, the deficiency was made up by Harold Kuhn (1953).

Kuhn's requirement of perfect recall is particularly easy to express in terms of possibility sets. One simply requires that any new information that Pandora receives results in a refinement of her knowledge partition—which means that each of her new possibility sets is always a subset of one of her previous possibility sets. Mini-Poker provides an example. In figure 8.3, Bob's later knowledge partition is a refinement of his earlier knowledge partition.

There is a continuing debate on how to cope with decision problems with imperfect recall that centers around the Absent-Minded Driver's problem. Terence must make two right turns to get to his destination.[4]

---

[4] Terence Gorman was the most absent-minded professor the world has ever seen.

He gets to a turning but can't remember whether he has already taken a right turn or not. What should he do?

Most attempts to answer the question take Bayesian decision theory for granted. But if Terence satisfies the omniscience requirement how could he have imperfect recall? Worse still, his possibility sets overlap, in flagrant contradiction of (Q2).

Knowledge versus belief. I think there is a strong case for distinguishing between knowledge and belief with probability one (section 8.5.2).

Mini-Poker again provides an example. In the left-hand game tree of figure 8.2, the branch of the tree that represents Alice's choice of raising after being dealt a *King* has been thickened to indicate that the players *believe* that, in equilibrium, she will take this action with probability one. In the right-hand game tree, her alternative choice of checking has been deleted to indicate that the players *know* from the outset that she will raise when holding a *King*.

The difference lies in the *reason* that the players act as though they are sure that Alice will raise when holding the *King*. In the knowledge case, she raises when holding a *King* in all possible worlds that are consistent with the model represented by the right-hand game tree. These possible worlds are all the available strategy profiles. Since the model rules out the possibility that Alice will check when holding a *King*, it is impossible that a rational analysis will predict that she will. In the belief case represented by the left-hand game tree, the set of all strategy profiles includes possibilities in which Alice checks when holding a *King*, but a rational analysis predicts that she won't actually do so.

In brief, when we know that Alice won't check, we don't consider the possibility that she might. When we don't know that she won't check, we consider the possibility that she might, but end up believing that she won't.

Alternative models. We don't need possibility sets to see that Alice is indifferent between checking and raising in Mini-Poker when holding a *Jack*. Bob plans to call a raise with probability $\frac{2}{3}$, and so Alice's expected payoff from raising is $\frac{2}{3} \times -2 + \frac{1}{3} \times 1 = -1$, which is the same as she gets from checking.

How would we obtain the same result using the apparatus of possibility sets? If we want to copy the method we employed for Bob, we can appeal to the new game tree for Mini-Poker shown in figure 8.4. In this formulation of the game, Bob decides whether to call or fold if Alice raises *before* she does anything at all. It doesn't matter when Bob makes this decision, as long as Alice doesn't discover what he has decided to do until *after* she makes her own decision.

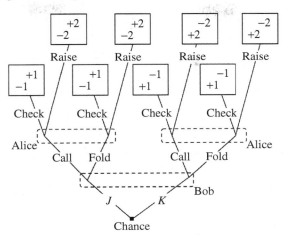

Figure 8.4. A different model for Mini-Poker. Bob decides in advance of Alice's move whether to call or fold if she raises. It doesn't matter when Bob makes this decision, as long as Alice doesn't discover what he has decided until after she makes her own decision.

When we ask how Alice's possibility sets evolve as the game proceeds, we now find that she has to consider a different state space than Bob, since our new model for Mini-Poker has eight possible plays instead of six. However, I won't pursue the analysis further because the point I want to make is already on the table.

If we were better disciplined, we would continually remind ourselves that everything we say is always relative to some model. When we forget to do so, we risk treating the underlying model as though it were set in concrete.

The game of Mini-Poker is only one of many examples that show that many alternative models are usually available. Much of the art of successful problem-solving lies in choosing a model from all these alternatives that is easy to analyze. It is for this reason that chapter 1 goes on at length about the advantages of formulating a model that satisfies Aesop's principle so that we can validly apply orthodox decision theory.

Metaphysicians prefer to think of their formalisms as attempts to represent the one-and-only true model, but I think such inflexible thinking is a major handicap in the kinds of problems that arise in ordinary life (section 8.5).

8.4 Knowledge in a Large World

Bayesian epistemology makes perfectly good sense in a small enough world, but we don't need to expand our horizons very far before we find ourselves in a large world. As soon as we seek to predict the behavior of other human beings, we should begin to be suspicious of discussions that take Bayesian decision theory for granted. After all, if Bob is as complex as Alice, it is impossible for her to create a model of the world that incorporates a model of Bob that is adequate for all purposes. Worse still, the kind of game-theoretic arguments that begin:

If I think that he thinks that I think ...

require that she contemplates models of the world that incorporate models of herself.

This section seeks to discredit the use of Bayesian epistemology in worlds in which such self-reference can't be avoided. I don't argue that it is *only* in such worlds that Bayesian epistemology is inappropriate. In most situations to which Bayesian decision theory is thoughtlessly applied—such as financial economics—the underlying world is so immensely large that nobody could possibly disagree with Savage that it is "utterly ridiculous" to imagine applying his massaging process to generate consistent beliefs (section 7.5.1). However, in worlds where self-reference is intrinsic, a knockdown argument against Bayesian epistemology is available.

I have to apologize to audiences at various seminars for claiming credit for this argument, but I only learned recently from Johan van Bentham that Kaplan and Montague (1960) proved a similar result more than forty years ago.

A possibility machine. How do rational people come to know things? What is the process of justification that philosophers have traditionally thought turns belief into knowledge? A fully satisfactory answer to this question would describe the process as an algorithm, in the manner first envisaged by Leibniz.

A modern Leibniz engine would not be a mechanical device of springs and cogs, but the program for a computer. We need not worry about what kind of computer or in what language the program is written. According to the Church–Turing hypothesis, all possible computations can be carried through on a highly simplified computer called a Turing machine.

Consider a hypothetical Leibniz engine L that may depend on whatever the true state of the world happens to be. The engine L decides what events Pandora will regard as possible under all possible circumstances. For each event E, it therefore determines the event $\mathcal{P}E$.

In this setup, we show that (P0) and (P2) are incompatible (figure 8.1). It follows that the same is true of (K0) and (K2). As Gödel showed for provability in arithmetic, we therefore can't hang on to both completeness and consistency in a large enough world when knowledge is algorithmic.

The Turing halting problem. We first need to be precise about how L decides possibility questions. It is assumed that L sometimes answers NO when asked suitably coded questions that begin:

> Is it possible that ...?

Issues of timing are obviously relevant here. How long should Pandora wait for an answer? Such timing problems are idealized away by assuming that she can wait for ever. If L never says NO, she regards it as possible that the answer is YES.

As in the Turing halting problem, we suppose that $[N]$ is the code for some question about the Turing machine N. We then take $\{M\}$ to be code for:

> Is it possible that M answers NO to $[M]$?

Let T be the Turing machine that outputs $\{x\}$ on receiving the input $[x]$. Now employ the Turing machine $I = LT$ that first runs an input through T, and then runs the output of T through the Leibniz engine L. Then the Turing machine I responds to $[M]$ as L responds to $\{M\}$.

An event E is now defined to be the set of states in which I responds to $[I]$ with NO. We then have the following equivalences:

$s \in E \Leftrightarrow I$ responds to $[I]$ with NO

$\qquad \Leftrightarrow L$ reports it to be impossible that I responds to $[I]$ with NO

$\qquad \Leftrightarrow s \in {\sim}\mathcal{P}E.$

It follows from (P2) that ${\sim}\mathcal{P}E = E \subseteq \mathcal{P}E$, which implies $\mathcal{P}E = B$. Thus, $E = {\sim}\mathcal{P}E = {\sim}B = \varnothing$, and so $\mathcal{P}\varnothing = B$. But (P0) says that $\mathcal{P}\varnothing = \varnothing$.

What is the moral? If the states in B are sufficiently widely construed and knowledge is algorithmic, then another way of reporting the preceding result is that we can't retain the infallibility assumption (K2) without abandoning (K0).

But we can't just throw away (K0) without also throwing away either (K1) or (K4).[5] It is unthinkable to abandon (K1), and so (K4) must go along with (K0). But (K4) is the property that bears the brunt of the requirement that all things have been considered in Pandora's world.

---

[5] It follows from (K1) that $E \subseteq F$ implies $\mathcal{K}E \subseteq \mathcal{K}F$. In particular, $\mathcal{K}\varnothing \subseteq \varnothing$, and so $\mathcal{K}\varnothing = \varnothing$. Thus ${\sim}\mathcal{K}{\sim}B = {\sim}\varnothing = B$. Therefore, $B = \mathcal{P}B = \mathcal{K}\mathcal{P}B = \mathcal{K}B$ by (K4). It is similarly easy to deduce (K3): $\mathcal{K}\mathcal{K}E = \mathcal{K}(\mathcal{P}\mathcal{K})E = (\mathcal{K}\mathcal{P})\mathcal{K}E = \mathcal{P}\mathcal{K}E = \mathcal{K}E.$

8.4.1   Algorithmic Properties?

The preceding impossibility result not only forces us to abandon either
$\mathcal{K}E \subseteq E$ or $\mathcal{K}B = B$, it also tells us that $\mathcal{K}E \subseteq E$ implies that $\mathcal{K}B = \varnothing$. One
can translate this property into the assertion that Pandora never knows
in what world she is operating. But what does this mean in practice?

One way of interpreting the result is to rethink what is involved in
saying that an event E has occurred. The standard view is that E is said
to occur if the true state s of the world has whatever property defines E.
But how do we determine whether s has the necessary property?

If we are committed to an algorithmic approach, we need an algorithmic procedure for the defining property P of each event E. This procedure can then be used to interrogate s with a view to getting a YES or NO
answer to the question:

Does s have property P?

We can then say that E has occurred when we get the answer YES and
that ~E has occurred when we get the answer NO.

But in a sufficiently large world, there will necessarily be properties for
which the relevant algorithm sometimes won't stop and give an answer,
but will rattle on forever. Our methodology will then fail to classify s
as belonging to either E or ~E. We won't then be able to implement
Ken Arrow's (1971, p. 45) ideal recipe for a state of the world—which
requires that it be described so completely that, if true and known, the
consequences of every action would be known (section 1.2).

In decision theory, we are particularly interested in questions like:

Is $\text{prob}(s) \leqslant p$?

Our algorithmic procedure might say YES for values of $p > \overline{p}$ and NO
for values of $p < \underline{p}$ but leave the question open for intermediate values
of p. So we have another reason for assessing events in large worlds in
terms of upper and lower probabilities.

8.4.2   Updating

We have just explored the implications of holding on to the infallibility
or consistency requirement (K2). However, my attempt in section 9.2 to
extend Bayesian decision theory to a minimal set of large worlds doesn't
take this line. I hold on instead to the completeness requirement (K0).

To see why, consider what happens when Pandora replaces her belief
space B by a new belief space F after receiving some information. In
Bayesian decision theory, we take for granted that $\mathcal{K}F = F$ so that F is

a truism in her new situation (section 8.3). If we didn't, it would take some fancy footwork to make sense of what is going on when we update prob(E) to prob($E \mid F$). In section 9.2, I therefore assume that Pandora never has to condition on events F that aren't truisms. To retain Bayes' rule, I go even further by insisting that any such F is measurable, and so can be assigned a unique probability.

For example, rational players in game theory are assumed to know that their opponents are also rational (section 8.5.3). As long as everybody behaves rationally and so play follows an equilibrium path, no inconsistency in what the players regard as known can occur. But rational players stay on the equilibrium path because of what *would* happen if they *were* to deviate. In the counterfactual world that would be created by such a deviation, the players would have to live with the fact that their knowledge that nobody will play irrationally has proved fallible. The more general case I consider in section 9.2 has the same character. It is only in counterfactual worlds that anything unorthodox is allowed—but it is what would happen if such counterfactual worlds were reached which ensures that they aren't reached.

8.5   Revealed Knowledge?

What are the implications of abandoning consistency (K2) in favor of completeness (K0)? If we were talking about knowledge of some absolute conception of truth, then I imagine everybody would agree that (K2) must be sacrosanct. But I plan to argue that absolute truth—whatever that may be—is irrelevant to rational decision theory. We can't dispense with some Tarskian notion of truth relative to a model, but as long as we don't get above ourselves by imagining we have some handle on the problem of scientific induction, then that seems to be all we need.

8.5.1   Jesting Pilate

Pontius Pilate has been vilified down the ages for asking:

What is truth?

His crime was to jest about a matter too serious for philosophical jokes. But is the answer to his question so obvious? Xenophanes didn't think so:

> But as for certain truth, no man knows it,
> And even if by chance he were to utter
> The final truth, he would himself not know it;
> For all is but a woven web of guesses.

Sir Arthur Eddington put the same point more strongly: "Not only is the universe stranger than we imagine, it is stranger than we can imagine." J. B. S. Haldane said the same thing in almost the same words. I too doubt whether human beings have the capacity to formulate a language adequate to describe whatever the ultimate reality may be by labeling some of its sentences as *true* and others as *false*. Some philosophers are scornful of such skepticism (Maxwell 1998). More commonly, they deny Quine's (1976) view that truth must be relative to a language. But how can any arguments they offer fail to beg the relevant questions?

In the absence of an ultimate model of possible realities, we make do in practice with a bunch of gimcrack models that everybody agrees are inadequate. We say that sentences within these models are true or false, even though we usually know that the entities between which relationships are asserted are mere fictions. As in quantum physics, we often tolerate models that are mutually contradictory because they seem to offer different insights in different contexts.

The game we play with the notion of truth when manipulating such models isn't at all satisfactory, but it seems to me the only genuine game in town. I therefore seek guidance in how to model knowledge and truth from the way these notions are actually used when making scientific decisions in real life, rather than appealing to metaphysical fancies of doubtful practicality.

8.5.2   Knowledge as Commitment to a Model

Bayesian decision theory attributes preferences and beliefs to Pandora on the basis of her choice behavior. We misleadingly say that her preferences and her beliefs are *revealed* by her choice behavior, as though we had reason to suppose her choices were really caused by preferences and beliefs. Perhaps Pandora actually does have preferences and beliefs, but all the theory of revealed preference entitles us to say is that Pandora makes decisions *as though* she were maximizing an expected utility function relative to a subjective probability measure.

Without denying that other interpretations may be more useful in other contexts, I advocate taking a similar revealed-preference approach to knowledge in the context of decision theory. The result represents a radical departure from the orthodox view of knowledge as justified true belief. With the new interpretation, knowledge need neither be justified nor true in the sense usually attributed to these terms. It won't even be classified as a particular kind of belief.

Revealed knowledge.   Instead of asking whether Pandora actually does know truths about the nature of the universe, I suggest that we ask instead what choice behavior on her part would lead us to regard her acting *as though* she knew some fact within her basic model.

To behave as though you know something is to act as though it were true. But we don't update truths like beliefs. If Pandora once knew that a proposition is true, we wouldn't allow her to say that she now knows that the same proposition is false without admitting that she was mistaken when she previously said the proposition was true. I therefore suggest saying that Pandora reveals that she knows something if she acts as though it were true in all possible worlds generated by her model.

In game theory, for example, the rules of a game are always taken to be common knowledge among the players (unless something is said to the contrary). The understanding is that, whatever the future behavior of the players, the rules will never be questioned. If Alice moves a bishop like a knight, we don't attempt to explain her maneuver by touting the possibility that we have misunderstood the rules of chess; we say that she cheated or made a mistake. In Bayesian decision theory, Pandora is similarly regarded as knowing the structure $D : A \times B \to C$ of her decision problem because she never varies this structure when considering the various possible worlds that may arise as a result of her choice of an action in A.

I think this attitude to knowledge is actually the *de facto* norm in scientific enquiry. Do electrons "really" exist? Physicists seem utterly uninterested in such metaphysical questions. Even when you are asked to "explain your ontology," nobody expects a disquisition on the ultimate nature of reality. You are simply being invited to clarify the working hypotheses built into your model.

Necessity. Philosophers say something is necessary if it is true in all possible worlds. The necessity operator $\Box$ and the corresponding possibility operator $\Diamond$ are then sometimes deemed to satisfy the requirements of figure 8.1. When knowledge is interpreted as commitment to a model, such arguments can be transferred immediately to the knowledge operator. However, there is a kick in the tail, because the arguments we have offered against assuming both (K0) and (K2) in a large world also militate against the same properties being attributed to the necessity operator.

Updating parameters. Bayesianism casts Bayes' rule in a leading role. But my separation of knowledge from belief allows Bayesian updating to be viewed simply as a humdrum updating of Pandora's beliefs about the parameters of a model whose structure she treats as known.

From this perspective, Bayesian updating has little or nothing to do with the problem of scientific induction. The kind of scientific revolution that Karl Popper (1959) studied arises when the data no longer allows Pandora to maintain her commitment to a particular model. She therefore throws away her old model and adopts a new model—freely admitting as she does so that she is being inconsistent.

It seems to me that the manner in which human beings manage this trick remains as much a mystery now as when David Hume first exposed the problem. There are Bayesians who imagine that prior probabilities could be assigned to all possible models of the universe, but this extreme form of Bayesianism has lost all contact with the reasons that Savage gave for restricting the application of his theory to small worlds. An even more extreme form of Bayesianism holds that scientists actually do use some form of Bayesian updating when revising a theory—but that they are unconscious of doing so. This latter notion has been put to me with great urgency on several occasions, but you might as well seek to persuade me that fairies live at the bottom of my garden.

Belief with probability one. How is belief with probability one to be distinguished from knowledge? On this subject, I share the views of Bayesians who insist that one should never allow subjective probabilities to go all the way to zero or one. Otherwise any attempt by Pandora to massage her system of beliefs into consistency would fail when she found herself trying to condition on a zero-probability event. This isn't to say that models in which events with extreme probabilities appear may not often be useful, but they should be regarded as limiting cases of models in which extreme probabilities don't appear (section 5.5.1).

8.5.3   Common Knowledge of Rationality

I want to conclude this chapter by returning to the issue of whether common knowledge of rationality implies backward induction (section 2.5.1). I have disputed this claim with Robert Aumann (Binmore 1996, 1997), but my arguments were based on the orthodox idea that we should treat common knowledge as the limiting case of common belief. In such a conceptual framework, Bob would find it necessary to update his belief about the probability that Alice will behave irrationally in the future if he finds that she has behaved irrationally in the past.

However, if knowledge is understood as being impervious to revision within a particular model (as I recommend in this section), then it becomes trivially true that common knowledge of rationality implies backward induction. No history of play can disturb the initial assumption that there is common knowledge of rationality, because knowledge isn't subject to revision. In predicting the behavior of the players in any subgame, we can therefore count on this hypothesis being available. It may be that some of these subgames couldn't be reached without Alice repeatedly choosing actions that game theorists would classify as irrational, but if Bob *knows* that Alice is rational, her history of irrational behavior is irrelevant. He will continue to act as though it were true that she is rational in all possible worlds that may arise.

9
Large Worlds

9.1 Complete Ignorance

There was once a flourishing literature on rational decision theory in large worlds. Luce and Raiffa (1957, chapter 13) refer to this literature as decision making under complete ignorance. They classify what we now call Bayesian decision theory as decision making under partial ignorance (because Pandora can't be completely ignorant if she is able to assign subjective probabilities to some events).

It says a lot about our academic culture that this literature should be all but forgotten. Presumably nobody reads Savage's (1951) *Foundations of Statistics* any more, since the latter half of the book is entirely devoted to his own attempt to contribute to the literature on decision making under complete ignorance.

The principle of insufficient reason. The problems with this attempt to harness orthodox probability theory to the problem of making decisions under complete ignorance were briefly reviewed in section 7.4.3. Figure 9.1 provides an example in which the principle of insufficient reason chooses either action a_1 or action a_2.

Critics allow themselves much freedom in constructing examples that discredit the principle, but a more modest version wouldn't be so easy to attack. For example, it is sometimes argued that our language—or the language in which we write our models—already incorporates implicit symmetries that would simply be made explicit if an appeal were made to a suitably restricted version of the principle of insufficient reason.

I am more enthusiastic about a much less ambitious proposal in which the class of traditional randomizing devices is expanded to other mechanisms for which it is possible to reject all reasons that might be given for regarding one outcome as more probable than another. One could then subsume such generalized lotteries into a theory of subjective probability in much the same way that traditional lotteries were subsumed in chapter 7. However, the need to explore all possible reasons for rejecting

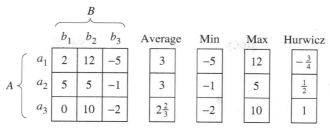

Figure 9.1. A decision problem. The diagram shows a specific decision problem $D : A \times B \rightarrow C$ represented as in figure 1.1 by a matrix. In this case, the consequences that appear in the matrix are Von Neumann and Morgenstern utilities. The ancillary columns are to help determine the action chosen by various decision criteria. The principle of insufficient reason chooses a_1 or a_2. The maximin criterion chooses a_2. The Hurwicz criterion with $h = \frac{1}{4}$ chooses a_3.

symmetry would seem incompatible with an assumption of complete ignorance.

The maximin criterion. We met the maximin criterion in section 3.7, where the idea that it represents nothing more than a limiting case of the Von Neumann and Morgenstern theory was denied. In figure 9.1, the maximin criterion chooses a_2.

A standard criticism of the maximin criterion is that it is too conservative. Suppose, for example, that Pandora is completely ignorant about how a number between 1 and 100 is to be selected. She now has to choose between two gambles, g_1 and g_2. In g_1, she wins $1m if the number 1 is selected and nothing otherwise. In g_2, she wins nothing if the number 100 is selected, and $1m otherwise. The gamble g_2 dominates g_1 because Pandora always gets at least as much from g_2 as g_1, and in some states of the world she gets more. But the maximin criterion tells her to be indifferent between the two gambles.

The Hurwicz criterion. Leo Hurwicz was a much-loved economist who lived a long and productive life, rounded off by the award of a Nobel prize for his work on mechanism design shortly before his death. He doubtless regarded the Hurwicz criterion as a mere *jeu d'esprit*. The criterion is a less conservative version of the maximin criterion in which Pandora evaluates an action by considering both its worst possible outcome and its best possible outcome. If c and C are the worst and the best payoffs she might get from choosing action a, Hurwicz (1951) suggested that Pandora should choose whichever action maximizes the value of

$$(1 - h)c + hC, \qquad (9.1)$$

where the Hurwicz constant h ($0 \leqslant h \leqslant 1$) measures how optimistic or pessimistic Pandora may be. In figure 9.1, the Hurwicz criterion with

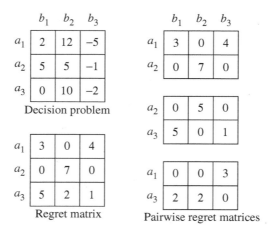

Figure 9.2. Minimax regret. The regret matrix is found by subtracting Pandora's actual payoff at an outcome from the maximum payoff she could have gotten in that state if she had chosen another action. She is then assumed to apply the minimax criterion to her regret matrix. The pairwise regret matrices are supplied to help check that Pandora's choices over pairs of actions reveal the intransitive preferences $a_1 \succ a_2 \sim a_3 \succ a_1$.

$h = \frac{1}{4}$ chooses a_3. With $h = 0$, it chooses a_2 because it then reduces to the maximin criterion.

A more sophisticated version of the Hurwicz criterion known as alpha-max-min has been proposed, but the generalization I propose later goes in another direction (Ghirardato et al. 2004; Marinacci 2002).

The minimax regret criterion. Savage's (1951) own proposal for making decisions under complete ignorance is the minimax regret criterion. Since he is one of our heroes, it is a pity that it isn't easier to find something to say in its favor.

The criterion first requires replacing Pandora's decision matrix of figure 9.1 by a regret matrix as in figure 9.2. The regrets are found by subtracting her actual payoff at an outcome from the maximum payoff she could have gotten in that state. She is then assumed to apply the minimax criterion (not the maximin criterion) to her regret matrix.[1]

One could take issue with Savage for choosing to measure regret by subtracting Von Neumann and Morgenstern utilities, but he could respond with an argument like that of section 4.3. However, the failure of the minimax regret criterion to honor Houthakker's axiom seems fatal (section 1.5.2). The example of figure 9.2 documents the failure.

---

[1] Much confusion can be avoided by remembering that economists usually evaluate consequences in terms of gains and so apply the maximin criterion. Statisticians usually evaluate consequences in terms of losses and so apply the minimax criterion.

The minimax regret criterion requires Pandora to choose a_1 from the set $A = \{a_1, a_2, a_3\}$. But when she uses the minimax regret criterion to choose from the set $B = \{a_1, a_3\}$, she chooses a_3.

9.1.1 Milnor's Axioms

John Milnor is famous for his classic book *Differential Topology*, but he was sufficiently interested in what his Princeton classmates like John Nash were doing that he wrote a paper on decision making under complete ignorance that characterizes all the criteria we have considered so far.

Milnor's analysis uses the matrix representation of a decision problem $D : A \times B \rightarrow C$ we first met in figure 1.1. The rows of the matrix represent actions in a feasible set A. The columns represent states of the world in the set B. As in figure 9.1, the consequences are taken to be Von Neumann and Morgenstern utilities.

For each such problem, Milnor assumes that Pandora has a preference relation $\preceq$ defined over the set A of actions. He then considers ten different axioms that we might impose on this preference relation. These ten axioms appear as the rows of the table in figure 9.3, which preserves the names that Milnor gave to them. The columns correspond to the four decision criteria considered earlier in this section. The circles show when a criterion satisfies a particular axiom. The stars indicate axioms that characterize a criterion when taken together.

Milnor's axioms are described below. Discussion of their significance is postponed until section 9.1.2.

1. **Ordering.** *Pandora has a full and transitive preference relation over the actions in A.*

2. **Symmetry.** *Pandora doesn't care how the actions or states are labeled.*

3. **Strong domination.** *If each entry in the row representing the action a_i strictly exceeds the corresponding entry in the row representing the action a_j, then Pandora strictly prefers a_i to a_j.*

4. **Continuity.** *Consider a sequence of decision problems with the same set of actions and states, in all of which $a_j \prec a_i$. If the sequence of matrices of consequences converges, then its limiting value defines a new decision problem, in which $a_j \preceq a_i$.*

5. **Linearity.** *Pandora doesn't change her preferences if we alter the zero and unit of the Von Neumann and Morgenstern utility scale used to describe the consequences.*

6. **Row adjunction.** *Pandora doesn't change her preferences between the old actions if a new action becomes available.*

	Principle of insufficient reason	Maximin criterion	Hurwicz criterion	Minimax regret criterion
Ordering	☆	☆	☆	☆
Symmetry	☆	☆	☆	☆
Strong domination	☆	☆	☆	☆
Continuity	○	☆	☆	☆
Linearity	○	○	☆	○
Row adjunction	☆	☆	☆	
Column linearity	☆			☆
Column duplication		☆	☆	☆
Convexity	○	☆		☆
Special row adjunction	○	○	○	☆

Figure 9.3. Milnor's axioms. This table is reproduced from Milnor's (1954) *Games against Nature*. The circles indicate properties that the criteria satisfy. The stars indicate properties that characterize a criterion when taken together.

7. Column linearity. *Pandora doesn't change her preferences if a constant is added to each consequence corresponding to a particular state.*

8. Column duplication. *Pandora doesn't change her preferences if a new column is appended that is identical to one of the old columns.*

9. Convexity. *Pandora likes mixtures. If she is indifferent between a_i and a_j and an action is available that always averages their consequences, then she likes the average action at least as much as a_i or a_j.*

10. Special row adjunction. *Pandora doesn't change her preferences between the old actions if a new action that is weakly dominated by all the old actions becomes available.*

Implications of column duplication. It seems to me that we can't do without axioms 1, 2, 3, 4, and 6, and so I shall call these the indispensable axioms. Savage's minimax regret criterion is therefore eliminated immediately because it fails to satisfy axiom 6 (row adjunction).

The indispensable axioms imply two conclusions that will be taken for granted in what follows. The first conclusion is that axiom 4 (continuity) can be used to generate a version of axiom 3 (strong domination) without the two occurrences of the word *strictly*. The second conclusion is that axioms 1, 2, and 6 imply that the order in which entries appear in any row of the consequence matrix can be altered without changing Pandora's preferences.

The axioms on which Milnor's work invites us to focus are axiom 7 (column linearity) and axiom 8 (column duplication). Taken together with the indispensable axioms, column linearity implies the principle of insufficient reason. Column linearity is therefore much more powerful than it may appear at first sight. But if we are completely ignorant about the states of the world, why should we acquiesce in a principle that tells us that a column which is repeated twice corresponds to two different but equiprobable states of the world? Perhaps it is really the same column written down twice. Taking this thought to the extreme, we are led to axiom 8 (column duplication).

Along with the indispensable axioms, column duplication implies that Pandora's preferences over actions depend only on the best and worst consequences in any row. To see why this is true, it is easiest to follow Milnor in identifying an action with the row of consequences to which it leads.

The two conclusions we have just derived from the indispensable axioms imply that

$$(c, c, \ldots, C) \preceq (c_1, c_2, \ldots, c_n) \preceq (C, C, \ldots, c), \qquad (9.2)$$

where the minimum of $c_1, c_2, \ldots, c_n$ is c and the maximum is C. We also need the second conclusion to infer that Pandora is indifferent between the two actions (c, C) and (C, c). It then follows that Pandora is indifferent between $(c, c, \ldots, C)$ and $(C, C, \ldots, c)$ because the first of the following matrices can be obtained from the second by column duplication:

$$\begin{bmatrix} c & c & \cdots & c & C \\ C & C & \cdots & C & c \end{bmatrix} \qquad \begin{bmatrix} c & C \\ C & c \end{bmatrix}.$$

The fact that Pandora is indifferent between all actions with the same best and worst outcomes can then be deduced from (9.2).

Characterizing the Hurwicz criterion. We have seen that the indispensable axioms plus column duplication imply that we need only consider actions of the form (c, C), where $c \leqslant C$. To characterize the Hurwicz criterion, we also need axiom 5 (linearity). Most commentators would include linearity among the indispensable axioms, but I think there is good reason to be more skeptical (section 9.1.2).

To obtain the Hurwicz criterion, Milnor begins by defining the Hurwicz constant h to be the smallest number no smaller than any k for which $(k, k) \preceq (0, 1)$. The continuity axiom then tells us that

$$(h, h) \sim (0, 1),$$

where $0 \leqslant h \leqslant 1$ (by strong domination). For $c < C$, the linearity axiom then implies that

$$((C - c)h + c, (C - c)h + c) \sim (c, C).$$

Finding the best (c, C) is therefore the same as maximizing the Hurwicz expression: $(1 - h)c + hC = (C - c)h + c$.

Characterizing maximin. I think that axiom 9 (convexity) is misplaced among Milnor's postulates. As with risk aversion, it tells us something about what Pandora happens to like, rather than how she must behave if she is to avoid irrationality. However, it suffices to ensure that $h = 0$ in the Hurwicz criterion.[2]

As previously, we need only consider actions of the form (c, C) with $c \leqslant C$. Milnor now applies the convexity axiom to the matrix

$$\begin{bmatrix} c & \frac{1}{2}(c + C) & \frac{1}{2}(c + C) \\ c & c & C \\ c & C & c \end{bmatrix} \tag{9.3}$$

to deduce that $(c, C) \preceq (c, \frac{1}{2}(c + C))$. By repeating this argument over and over again with C replaced by $\frac{1}{2}(c + C)$ and then appealing to axiom 4 (continuity), we are led to the conclusion that $(c, C) \preceq (c, c)$. Weak domination then yields that $(c, C) \sim (c, c)$. Finding the best (c, C) is therefore the same as maximizing c.

9.1.2 Discussion of Milnor's Axioms

Milnor's axioms treat Pandora's preferences over actions as fundamental, but it is easy to recycle the arguments of chapter 1 so that his results can be absorbed into the theory of revealed preference. We simply treat Pandora's choices between acts in the set $\aleph$ of all acts as fundamental, and then use the preferences she reveals when these choices are consistent as the necessary input into Milnor's theory. His axiom 6 (row adjunction) can then be replaced by Houtthaker's axiom, which allows us to dispense with the need to assume transitivity of preference in his axiom 1 (ordering).

Milnor's identification of the final consequences in a decision problem with Von Neumann and Morgenstern utilities also needs some consideration. He thereby implicitly assumes a version of our postulate 3, which allows a prize in a lottery to be replaced by any independent prize that Pandora likes equally well (section 3.4.1). We extended this postulate to

---

[2] Milnor's characterization of the maximin criterion is only one of many. Perhaps the most interesting alternative is Barbara and Jackson (1988).

gambles when developing Bayesian decision theory in section 7.2, and there seems no good reason for accepting it there but not here. In any case, section 6.4.1 has already signaled my intention of upholding this assumption even in the case of totally muddling boxes.

Linearity. Luce and Raiffa (1957, chapter 13) provide an excellent discussion of the background to Milnor's axioms that is partly based on a commentary by Herman Chernov (1954). However, neither in their discussion nor anywhere else (as far as I know) is any serious challenge offered to Milnor's axiom 5 (linearity), which says that we are always free to alter the zero and the unit on Pandora's Von Neumann and Morgenstern utility scale without altering her preferences among actions. It is for this reason that time was taken out in section 3.6.2 to explain the circumstances under which it makes sense to identify an absolute zero on Pandora's utility scale.

My reasons for arguing that the existence of an absolute zero should be treated as the default case become more pressing in a large-world context, and I therefore think it a mistake to insist on axiom 5 when contemplating decisions made under complete ignorance. But abandoning axiom 5 rules out all four of the decision criteria that appear in Milnor's table of figure 9.3. However, we shall find that the Hurwicz criterion can be resuscitated in a multiplicative form rather than the classical additive form.

Separating preferences and beliefs? With the indispensable axioms, Milnor's axiom 7 (column linearity) implies the principle of insufficient reason. The reasons for rejecting linearity apply with even greater force to column linearity, but abandoning axiom 7 entails a major sacrifice. Luce and Raiffa (1957, p. 290) make this point by noting that axiom 7 follows from Rubin's axiom, which says that Pandora's preferences over actions should remain unaltered if some random event after her choice of action might replace her original decision problem by a new decision problem whose consequence matrix has rows that are all the same, so that her choice of action is irrelevant.

Denying Rubin's axiom as a general principle puts paid to any hopes we might have had of sustaining the purity of Aesop's principle in a large world. The probability with which Pandora's choice doesn't matter would make no difference to her optimal choice if her preferences and beliefs were fully separated. We therefore have to learn to tolerate some muddling of the boundary between preferences and beliefs in a large world. A major virtue of Milnor's axiom system is that it forces us to confront such inconsistencies between our different intuitions—rather as the Banach–Tarski paradox forces us to face up to similar inconsistencies in our intuitions about the nature of congruent sets in Euclidean space (section 5.2.1).

Taking account only of maxima and minima. We have seen that it is Milnor's axiom 8 (column duplication) which allows us to escape being trapped in the principle of insufficient reason. It is easy to respond that column duplication is too strong an assumption, but I think such criticism overlooks Milnor's insistence that he really is talking about making decisions under *complete* ignorance. On this subject, Milnor (1954, p. 49) says:

> Our basic assumption that the player has absolutely no information about Nature may seem too restrictive. However such no-information games may be used as a normal form for a wider class of games in which certain types of partial information are allowed. For example if the information consists of bounds for the probabilities of the various states of Nature, then by considering only those mixed strategies for Nature which satisfy these bounds, we construct a new game having no information.

My approach in the rest of this chapter can be seen partly as an attempt to make good on this prognosis, with upper and lower probabilities serving as the bounds on the probabilities of the states of nature to which Milnor refers. In taking this line, I follow Klibanoff, Marinacci, and Mukerji (2005), who derive a version of the Hurwicz criterion by modifying the Von Neumann and Morgenstern postulates. However, I am more radical, since I abandon Milnor's linearity axiom, which allows me to work with products of utilities rather than sums.

Convexity. I have already observed that Milnor's axiom 9 (convexity) doesn't belong in a list of rationality axioms. However, if one is going to make this kind of assumption it seems to me that one can equally well argue in favor of anti-convexity:[3]

9*. Anti-convexity. *Pandora regards a mixture of two actions as optimal if and only if she also regards each of the two actions as optimal as well.*

What happens to Milnor's defense of the maximin criterion if we replace his axiom 9 by the new axiom 9*? Begin by switching the roles of c and C in the matrix (9.3). If the first row is optimal in the new matrix, then anti-convexity implies that the second and third rows are optimal too. If the first row isn't optimal, then either the second or the third row is optimal. Either way, we find that $(C, c) \succeq (C, \frac{1}{2}(C + c))$. Milnor's argument then yields that $(C, c) \succeq (C, C)$ and so we end up with the maximax criterion in place of the minimax criterion!

---

[3] Axiom 9* follows from Rubin's axiom, symmetry, and row adjunction.

Something has obviously gone badly wrong here. We can even generate a contradiction by writing down the most natural version of axiom 9:

9**. Indifference to mixtures. *If Pandora is indifferent between two actions, then she is indifferent between all mixtures of the actions.*

This axiom is inconsistent with column duplication and the indispensable axioms, because it simultaneously implies both the maximin and maximax criteria.

I think our intuitions are led astray by pushing the analogy with lotteries too far. In particular, I don't feel safe in making any a priori assumptions that say how Pandora should value mixtures of muddling boxes in terms of the muddling boxes that get mixed (section 6.4). To do so is to proceed as though nonmeasurable sets behave just like measurable sets.

9.2 Extending Bayesian Decision Theory

If Pandora isn't ignorant at all, then she knows which state of the world applies in advance of making any decision. If she is completely ignorant, she decides without any information at all. There are many intermediate possibilities corresponding to all the different items of partial information that may be available to her—usually in a vague form that is difficult to formalize. Bayesianism holds that all such cases of partial ignorance can be dealt with by assigning unique subjective probabilities to the states of the world. I think that Bayesian decision theory only applies in small worlds in which Pandora's level of ignorance can be regarded as being relatively low (section 7.1.1). So there is a gap waiting to be filled between the case of low partial ignorance and the case of total ignorance.

I now follow numerous others—notably Gilboa (2004) and Gilboa and Schmeidler (2001)—in trying to fill some of this gap, but without following the usual practice of proposing a system of axioms or postulates. This wouldn't be hard to do, but I feel that the axiom–theorem–proof format stifles the kind of open-minded debate that is appropriate in such a speculative enterprise.

A minimal extension of Bayesian decision theory. I have no idea how to proceed if one isn't allowed to appeal to Savage's massaging process, so that Pandora can be supposed to have attained some level of consistency in her subjective beliefs (section 7.5.1). The large worlds to which my theory applies are therefore not the immensely large worlds that would need to be considered if we were aspiring to solve the problem of scientific induction. My worlds go beyond the Bayesian paradigm only to the extent that Pandora isn't required to produce a determinate subjective probability for each and every event.

When massaging her initial snap judgments, Pandora may find that an event E about which she is ignorant must be assigned an upper probability $\overline{p}(E) < 1$ and a lower probability $\underline{p}(E) > 0$ for consistency reasons, but we don't require that she tie her beliefs down any further (section 5.6). If she is completely ignorant, then $\overline{p}(E) = 1$ and $\underline{p}(E) = 0$.

One may argue that features of the event E that aren't strictly probabilistic in character may also be relevant, so that factors other than $\overline{p}(E)$ and $\underline{p}(E)$ need to be taken into account in evaluating E. It is partly for this reason that muddling boxes were introduced in section 6.5. They are intended to provide an example of a case in which only the upper and lower probabilities of an event have any significance. Just as roulette wheels exemplify the kind of mechanism to which the models of classical probability theory apply, muddling boxes are intended to exemplify the kind of mechanism to which the calculus of upper and lower probabilities applies.

We saw in section 5.6.3 that Giron and Rios (1980) examined the extent to which Bayesian decision theory can be applied in such a situation. However, in their quasi-Bayesian theory, Pandora doesn't have a full set of preferences over the set $\aleph$ of acts. My theory goes further by postulating that she does have a full set of preferences over acts, even though her beliefs are incomplete in the Bayesian sense.

A price has to be paid for such an extension of the theory: we can no longer ask that Pandora's beliefs over states in the set B can be fully separated from her preferences over consequences in the set C. One way to say this is that we follow Milnor in sacrificing his column linearity for column duplication (section 9.1.2).

9.2.1 Evaluating Nonmeasurable Events

In the following gamble G, Pandora gets the best possible prize $\mathcal{W}$ if the event E occurs and the worst possible prize $\mathcal{L}$ if the complementary event $\sim E$ occurs:

$$G \;=\; \begin{array}{|c|c|} \hline \mathcal{L} & \mathcal{W} \\ \hline \sim E & E \\ \hline \end{array}. \qquad\qquad (9.4)$$

In section 7.2, we made assumptions that allowed a probability $p(E)$ to be assigned to the event E, but we are now concerned with the case in which the event E isn't measurable. However, other events are assumed to be measurable, so that it becomes meaningful to talk about the upper probability $P = \overline{p}(E)$ and the lower probability $p = \underline{p}(E)$ of the event E.

If we assume that the Von Neumann and Morgenstern postulates apply when the prizes in lotteries are allowed to be gambles of any kind, then we can describe Pandora's choice behavior over such lotteries with a Von Neumann and Morgenstern utility function u, which we can normalize so that $u(L) = 0$ and $u(W) = 1$.

The next step is then to accept the principle that motivates Milnor's column duplication axiom by assuming that Pandora's utility for the gamble G of (9.4) depends only on the upper and lower probabilities of E, so that we can write

$$u(G) = U(p, P). \tag{9.5}$$

Quasi-probability. Our normalization of the Von Neumann and Morgenstern utility function u makes it possible to regard $u(G)$ not only as Pandora's utility for a particular kind of gamble, but also as a kind of quasi-probability of the event E. When doing so, I shall write

$$u(G) = \pi(E). \tag{9.6}$$

Although the assumptions to be made will result in π turning out to be nonadditive, I shan't call it a nonadditive probability as is common in the literature for analogous notions (section 5.6.3). I think this would be too misleading for a theory that doesn't involve computing generalized expected values of any kind (whether using the Choquet integral or not). To evaluate a gamble like

$$\begin{array}{|c|c|c|c|c|}
\hline
\mathcal{P}_1 & \mathcal{P}_2 & \mathcal{P}_3 & \cdots & \mathcal{P}_n \\
\hline
E_1 & E_2 & E_3 & \cdots & E_n \\
\hline
\end{array}, \tag{9.7}$$

Pandora will be assumed to begin by replacing each consequence $\mathcal{P}_i$ by an independent win-or-lose lottery that she likes equally well. She then works out the event F in which she will get the best prize W in the resulting compound gamble. Her final step is to choose a feasible gamble that maximizes the value of

$$U(\underline{p}(F), \overline{p}(F)).$$

These remarks aren't meant to imply that the various theories of non-additive probability that have been proposed have no applications. A theory that dealt adequately with all the ways in which Pandora might be partially ignorant would presumably need to be substantially more complicated than the theory I am proposing here.

9.2.2 The Multiplicative Hurwicz Criterion

What properties should be attributed to the function U? The following assumptions seem uncontroversial:

(1) $p = U(p, p) \leqslant U(p, P) \leqslant u(P, P) = P$;

(2) $U(p, P) \leqslant U(p, Q) \quad \Leftrightarrow \quad p \leqslant P \leqslant Q$;

(3) $U(p, P) \leqslant U(q, P) \quad \Leftrightarrow \quad p \leqslant q \leqslant P$.

The second and third assumptions are reminiscent of the separable preferences discussed in section 3.6.2. The analogy can be pushed much further by considering Pandora's attitude to independent events.

The rest of this section argues that we can thereby justify the use of a multiplicative Hurwicz criterion in which U takes the form

$$U(p, P) = p^{1-h} P^h$$

The constant h ($0 \leqslant h \leqslant 1$) will be called a (multiplicative) Hurwicz coefficient.

Independence. If a roulette wheel is spun twice, everybody agrees that the slot in which the little ball stops on the first spin is independent of the slot on which it stops on the second spin. If s_1 is the slot that results from the first spin and s_2 is the slot that results from the second spin, then we write (s_1, s_2) to represent the state of the world in which both spins are considered together. The set of all such pairs of states is denoted by $S_1 \times S_2$, where S_1 is the set of all states than can arise at the first spin and S_2 is the set of all states that can arise at the second spin.

More generally, suppose that $B = B_1 \times B_2$, where it is understood that all events E in B_1 are independent of all events F in B_2. The plan is to look at Pandora's attitude to the set of all events $E \times F$. We suppose that

(1) $(L_F, M_E) \prec (L_F, M_{E'})$ implies $(L_{F'}, M_E) \preceq (L_{F'}, M_{E'})$,

(2) $(L_F, M_E) \prec (L_{F'}, M_E)$ implies $(L_F, M_{E'}) \preceq (L_{F'}, M_{E'})$,

where the lotteries L_F and M_E are of the type indicated in figure 9.4.

We can now recycle the argument of section 3.6.2 to obtain an equation analogous to (3.3):

$$\pi(E \times F) = \pi_1(E)\pi_2(F) + A\pi_1(E)(1 - \pi_2(F)) + B\pi_2(F)(1 - \pi_1(E)).$$

Since $A = \pi(B_1, \varnothing) = 0$ and $B = \pi(\varnothing, B_2)$, we obtain that

$$\pi(E \times F) = \pi(E)\pi(F), \tag{9.8}$$

where the events E and F have been identified with $E \times B_2$ and $B_1 \times F$.

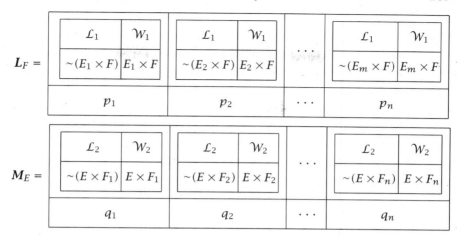

Figure 9.4. Some lotteries. The prizes in lotteries of type L_F incorporate different E_i but always have the same F. The prizes in lotteries of type M_E incorporate different F_j but always have the same E.

To go further, we rewrite (9.8) in terms of the function U. For this purpose, take $p = \underline{p}(E)$, $P = \overline{p}(E)$, $q = \underline{p}(F)$, and $Q = \overline{p}(F)$. But $\underline{p}(E \times F) = pq$ and $\overline{p}(E \times F) = PQ$, and so we are led to the functional equation

$$U(pq, PQ) = U(p, P)U(q, Q).$$

Such functional equations are hard to satisfy. The only continuously differentiable solutions take the multiplicative form

$$U(p, P) = p^{1-h}P^h \tag{9.9}$$

that we have been seeking to establish (section 10.6).

Additivity? Because $\underline{p}(E) = 1 - \overline{p}(\sim E)$ (section 5.6.2), it is always true that

$$\underline{p}(E) + \underline{p}(\sim E) \leqslant 1 \quad \text{and} \quad \overline{p}(E) + \overline{p}(\sim E) \geqslant 1$$

with equality only in the measurable case when $\underline{p}(E) = \overline{p}(E)$. More generally,

$$\pi(E) + \pi(\sim E) \leqslant 1,$$

for all events E if and only if $h \leqslant \frac{1}{2}$ (section 10.7). If $h \leqslant \frac{1}{2}$, then $\pi(E) + \pi(\sim E) = 1$ if and only if $\underline{p}(E) = \overline{p}(E)$.

The case when $h \geqslant \frac{1}{2}$ isn't symmetric because the inequality of the arithmetic and geometric means says that

$$p^{1-h}P^h \leqslant (1-h)p + hP$$

with equality if and only if $p = P$.

Absolute zero. If $\underline{p}(E) = 0$, then $\pi(E) = 0$ no matter what the value of $\overline{p}(E)$ may be. It is for this reason that the case of an absolute zero was discussed in section 3.6.2. However, in everyday decision problems, the existence of an absolute zero is unlikely to upset any standard intuitions. The following application of the multiplicative Hurwicz criterion to the Ellsberg paradox provides an example.

9.2.3 Uncertainty Aversion

The Ellsberg paradox is explained in figure 5.2. Subjects in laboratory experiments commonly reveal the preferences

$$J \succ K \quad \text{and} \quad L \succ M,$$

but there isn't any way of assigning subjective probabilities to the variously colored balls that makes these choices consistent with maximizing expected utility. Such behavior is usually explained by saying that most people are averse to ambiguity or uncertainty in a manner that is incompatible with Bayesian decision theory.[4] However, the behavior is compatible with the minimal extension considered here.

To see why, suppose that Pandora is indifferent between winning one million dollars in an Ellsberg lottery and participating in a lottery in which she gets her best possible outcome $\mathcal{W}$ with probability b and her worst possible outcome $\mathcal{L}$ with probability $1 - b$. Suppose she is also indifferent between losing and participating in a lottery in which she gets $\mathcal{W}$ with probability s and $\mathcal{L}$ with probability $1 - s$, where $b > s$. If $\epsilon > 0$ is small and $b/s = 1 + \epsilon$, then

$$u(J) = \tfrac{2}{3}s + \tfrac{1}{3}b = s(1 + \tfrac{1}{3}\epsilon),$$
$$u(K) = s^{1-h}(\tfrac{1}{3}s + \tfrac{2}{3}b)^h = s(1 + \tfrac{2}{3}\epsilon)^h \approx s(1 + \tfrac{2}{3}\epsilon h),$$
$$u(L) = \tfrac{1}{3}s + \tfrac{2}{3}b = s(1 + \tfrac{2}{3}\epsilon),$$
$$u(M) = (\tfrac{2}{3}s + \tfrac{1}{3}b)^{1-h}b^h = s(1 + \tfrac{1}{3}\epsilon)^{1-h}(1 + \epsilon)^h \approx s(1 + (\tfrac{1}{3} + \tfrac{2}{3}h)\epsilon).$$

In everyday decision problems, Pandora will therefore reveal preferences that are uncertainty averse whenever the Hurwicz coefficient $h < \tfrac{1}{2}$. For this reason, I think that the case when $h > \tfrac{1}{2}$ is only of limited interest.

9.2.4 Bayesian Updating

How does one update upper and lower probabilities after receiving a new piece of information (Klibanoff and Hanany 2007)? There seems to be nothing approaching a consensus in the literature on a simple formula

---

[4] See, for example, Dow and Werlang (1992), Epstein (1999, 2001), and Ryan (2002).

that relates the prior upper and lower probabilities with the posterior upper and lower probabilities. Observing a new piece of information may even lead Pandora to expand her class of measurable sets so that all her subjective probabilities need to be recalculated from scratch.

However, if we restrict our attention to the special case discussed in section 8.4.2, matters become very simple. If Pandora never has to condition on events F other than truisms to which she can attach a unique probability, then we can write

$$\underline{p}(E \mid F)p(F) = \underline{p}(E \text{ and } F), \qquad \overline{p}(E \mid F)p(F) = \overline{p}(E \text{ and } F).$$

9.3 Muddled Strategies in Game Theory

Anyone who proposes a decision theory that distinguishes between risk and uncertainty naturally considers the possibility that it might have interesting applications in game theory.[5] Because of the way my own theory treats independence, it is especially well adapted for this purpose.

Mixed strategies. If a game has a rational solution, it must be one of the game's Nash equilibria (section 2.2). In figure 2.1, the cells of the payoff tables in which both payoffs are circled correspond to Nash equilibria, because each player is then making a best reply to the strategy choice of the other. However, in the game Matching Pennies of figure 9.5, none of the cells have both their payoffs circled. It follows that Matching Pennies has no Nash equilibrium in pure strategies.

To solve games like Matching Pennies, we need to extend the class of pure strategies to a wider class of mixed strategies—an idea of Von Neumann that was anticipated by the mathematician Émile Borel. A mixed strategy requires that players use a random device to choose one of their pure strategies with predetermined probabilities. For example, if Alice bluffs with probability $\frac{1}{3}$ when holding the Jack in Mini-Poker, she is using a mixed strategy (section 8.3.1).

In Matching Pennies, it is a Nash equilibrium if each player uses the mixed strategy in which *heads* and *tails* are played with equal probability. They are then using the maximin strategy that Von Neumann identified as the rational solution of such two-person zero-sum games.

John Nash showed that all finite games have at least one Nash equilibrium when mixed strategies are allowed, but his result doesn't imply that the problem of identifying a rational solution of an arbitrary game is

---

[5] See, for example, Dow and Werlang (1994), Greenberg (2000), Lo (1996, 1999), Kelsey and Eichberger (2000), Marinacci (2000), and Mukerji and Shin (2002).

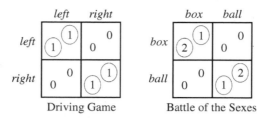

Driving Game   Battle of the Sexes

Figure 9.5. Two games. Matching Pennies has a mixed Nash equilibrium in which Alice and Bob both choose *heads* or *tails* with probability $\frac{1}{2}$. The Battle of the Sexes has two Nash equilibria in pure strategies. Like all symmetric games, it also has a symmetric equilibrium in which both players use their second pure strategy with probability $\frac{1}{3}$. At this equilibrium each player gets only their security level. However, to guarantee their security levels, each player must use their second pure strategy with probability $\frac{2}{3}$.

solved. Aside from other considerations, most games have many equilibria. Deciding which equilibrium will be played by rational players is one aspect of what game theorists call the equilibrium selection problem.

Battle of the Sexes. The Battle of the Sexes of figure 9.5 is an example invented by Harold Kuhn to illustrate how intractable the equilibrium selection problem can be, even in very simple cases.[6]

The politically incorrect story that goes with this game envisages that Adam and Eve are honeymooning in New York. At breakfast, they discuss where they should meet up if they get separated during the day. Adam suggests that they plan to meet at that evening's boxing match. Eve suggests a performance of *Swan Lake*. Rather than spoil their honeymoon with an argument, they leave the question unsettled. So when they later get separated in the crowds, each has to decide independently whether to go to the boxing match or the ballet. The Battle of the Sexes represents their joint dilemma.

The circled payoffs in figure 9.5 show that the game has two Nash equilibria in pure strategies: (*box, box*) and (*ball, ball*). All symmetric games have at least one symmetric Nash equilibrium when mixed strategies are allowed. The Battle of the Sexes has a unique symmetric equilibrium, whose properties are made evident by the right-hand diagram of figure 9.6.

The horizontal chord shows all payoff pairs that are possible if Adam plays his second pure strategy with probability $p = \frac{1}{3}$. All of Eve's strategies are best replies to this choice of mixed strategy by Adam. In particular, it is a best reply if she forces the outcome of the game to lie on

---

[6] For a more extensive discussion of the classical game, see Luce and Raiffa (1957, p. 90).

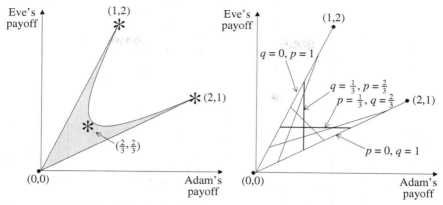

Figure 9.6. Battle of the Sexes. The very nonconvex set shaded in the left-hand diagram is the noncooperative payoff region of the Battle of the Sexes. It consists of all payoff pairs that can be achieved if Adam and Eve choose their mixed strategies independently. The stars show the payoff pairs at the three Nash equilibria of the game. The chords in the right diagram can be thought of as representing the players' mixed strategies. Those drawn consist of all multiples of $p = \frac{1}{6}$ and $q = \frac{1}{6}$, where p is the probability with which Adam chooses *ball*, and q is the probability with which Eve chooses *box*. As p takes all values between 0 and 1, the corresponding chord sweeps out the game's noncooperative payoff region. The mixed Nash equilibrium of the game occurs where the horizontal chord corresponding to $p = \frac{1}{3}$ crosses the vertical chord corresponding to $q = \frac{1}{3}$.

the vertical chord of figure 9.6 by playing her second pure strategy with probability $q = \frac{1}{3}$. All of Adam's strategies are best replies to this mixed strategy, including $p = \frac{1}{3}$. We have therefore found a mixed Nash equilibrium in which Adam plays $p = \frac{1}{3}$ and Eve plays $q = \frac{1}{3}$. In this symmetric equilibrium, each player ends up with an expected payoff of $\frac{2}{3}$.

Solving the Battle of the Sexes? What is the rational solution of the Battle of the Sexes? If the question has an answer, it must be one of the game's Nash equilibria. If we ask for a solution that depends only on the strategic structure of the game, then the solution of a symmetric game like the Battle of the Sexes must be a symmetric Nash equilibrium.[7] Can we therefore say that the rational solution of the Battle of the Sexes requires that each player should use their second pure strategy with probability $\frac{1}{3}$?

The problem with this attempt to solve the Battle of the Sexes is that it gives both players an expected payoff of no more than $\frac{2}{3}$, which also happens to be their maximin payoff—the largest expected utility a player can

---

[7] The Driving Game is solved in the United States by choosing the asymmetric Nash equilibrium in which everybody drives on the right. This solution depends on the context in which the game is played, and not just on the strategic structure of the game. For example, the game is solved in Japan by having everybody drive on the left.

guarantee independently of the strategy used by the other player. However, Adam and Eve's maximin strategies aren't the same as their equilibrium strategies. To guarantee an expected payoff of $\frac{2}{3}$, Adam must choose $p = \frac{2}{3}$, which makes the corresponding chord vertical in figure 9.6. Eve must choose $q = \frac{2}{3}$, which makes the corresponding chord horizontal.

This consideration undercuts the argument offered in favor of the symmetric Nash equilibrium as the solution of the game. If Adam and Eve are only expecting an expected payoff of $\frac{2}{3}$ from playing the symmetric Nash equilibrium, why don't they switch to playing their maximin strategies instead? The payoff of $\frac{2}{3}$ they are expecting would then be *guaranteed*. At one time, John Harsanyi (1964, 1966) argued that the rational solution of the Battle of the Sexes therefore requires that Adam and Eve should play their maximin strategies.

Harsanyi's suggestion would make good sense if the Battle of the Sexes were a zero-sum game like Matching Pennies, since it would then be a Nash equilibrium for Adam and Eve to play their maximin strategies. But it isn't a Nash equilibrium for them to play their maximin strategies in the Battle of the Sexes. For example, if Adam plays $p = \frac{2}{3}$, Eve's best reply is $q = 0$.

The problem isn't unique to the Battle of the Sexes (Aumann and Maschler 1972). If we are willing to face the extra complexity posed by asymmetric games, we can even write down 2×2 examples with a unique Nash equilibrium in which one of the players gets no more than his or her maximin payoff.

My plan is to use this old problem to illustrate why it can sometimes be useful to expand the class of mixed strategies to a wider class of muddled strategies in the same kind of way that Borel expanded the class of pure strategies to the class of mixed strategies.

9.3.1 Muddled Strategies

To implement a mixed strategy, players are assumed to use a randomizing device like a roulette wheel or a deck of cards. To implement a muddled strategy, they use muddling boxes (section 6.5).

In the Battle of the Sexes, a muddled strategy for Adam is therefore described by specifying the lower probability p and the upper probability P with which he will play his second pure strategy. A muddled strategy for Eve is similarly described by specifying the upper probability Q and the lower probability q with which she will play her second pure strategy.

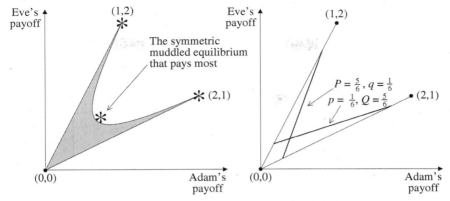

Figure 9.7. A muddled equilibrium in the Battle of the Sexes. In the case $h = \frac{1}{2}$, there is a symmetric Nash equilibrium in which each player uses a muddled strategy with lower probability approximately $\frac{1}{6}$ and upper probability approximately $\frac{5}{6}$. The equilibrium payoff to each player exceeds $\frac{3}{4}$.

When muddled strategies are used in the Battle of the Sexes, I shall assume that the players' preferences are determined by the multiplicative Hurwicz criterion (section 9.2.2). The Hurwicz coefficient h will be assumed to be the same for both players in order to keep everything symmetric. It will also be assumed that neither player likes uncertainty, so $0 \leqslant h \leqslant \frac{1}{2}$ (section 9.2.3). We will get nothing new if $h = 0$, and so the case when the Hurwicz criterion reduces to the maximin criterion is excluded as well. Under these conditions, section 10.7 proves a number of results. In particular:

> The Battle of the Sexes doesn't just have one symmetric
> Nash equilibrium; it has a continuum of symmetric
> equilibria if muddled strategies are allowed.

Figure 9.7 illustrates the case when $h = \frac{1}{2}$. The symmetric equilibrium that pays most then requires that both players use a muddled strategy whose upper probability is approximately $\frac{5}{6}$ and whose lower probability is approximately $\frac{1}{6}$. The payoff to each player at this equilibrium exceeds $\frac{3}{4}$, and hence lies outside the noncooperative payoff region when only mixed strategies are allowed. In particular, both players get more than their maximin payoff of $\frac{2}{3}$ at this symmetric equilibrium.

It would be nice if this last result were true whenever $h \leqslant \frac{1}{2}$, but it holds only for values of h sufficiently close to $\frac{1}{2}$. So the general problem of finding a rational solution to the Battle of the Sexes in a symmetric environment remains open.

9.4 Conclusion

This book follows Savage in arguing that Bayesian decision theory is applicable only in worlds much smaller than those in which it is standardly applied in economics. This chapter describes my own attempt to extend the theory to somewhat larger worlds. Although the extension is minimal in character, it nevertheless turns out that even very simple games can have new Nash equilibria if the players use the muddled strategies that my theory allows. I don't know how important this result is for game theory in general. Nor do I know to what extent the larger worlds to which my theory applies are an adequate representation of the economic realities to which Bayesian decision theory has been applied in the past. Perhaps others will explore these questions, but I fear that Bayesianism will roll on regardless.

10
Mathematical Notes

10.1 Compatible Preferences

This section justifies the representation of separable preferences of section 3.6.2.

Two preference relations $\preceq_1$ and $\preceq_2$ will be said to be compatible if it is always true that

$$a \prec_1 b \quad \text{implies} \quad a \preceq_2 b.$$

(It then follows that $a \prec_2 b$ implies $a \preceq_1 b$.) If the two preference relations are defined on the set $\text{lott}(C)$ of lotteries with prizes in C and satisfy the Von Neumann and Morgenstern postulates, then they may be described with Von Neumann and Morgenstern utility functions $v_1 : C \rightarrow \mathbb{R}$ and $v_2 : C \rightarrow \mathbb{R}$. If they are also compatible on $\text{lott}(C)$, then $v_1 = Av_2 + B$ or $v_2 = Av_1 + B$, where $A \geqslant 0$ and B are constants. (If $A > 0$, then $\preceq_1$ and $\preceq_2$ are the same preference relation. If $A = 0$, at least one of $\preceq_1$ and $\preceq_2$ is complete indifference.)

With these preliminaries in place, we can return to section 3.6.2 to establish the representation (3.3) for a separable preference relation $\preceq$ on $\text{lott}(C)$. Begin by defining the preference relation $\preceq_L$ on $\text{lott}(C_2)$ by

$$M \preceq_L M' \quad \Leftrightarrow \quad (L, M) \preceq (L, M')$$

and define $\preceq_M$ on $\text{lott}(C_1)$ similarly.

If trivial cases are excluded, we can find y_1 in C_1 and y_2 in C_2 so that $\preceq_{y_1}$ and $\preceq_{y_2}$ aren't total indifference relations. Normalize the Von Neumann and Morgenstern utility function u that represents $\preceq$ on $\text{lott}(C)$ so that $u(y_1, y_2) = 0$, and define $u_1 : C_1 \rightarrow \mathbb{R}$ and $u_2 : C_2 \rightarrow \mathbb{R}$ by

$$u_1(c_1) = u(c_1, y_2) \quad \text{and} \quad u_2(c_2) = u(y_1, c_2).$$

Since $\preceq_L$ is compatible with $\preceq_{y_1}$ and $\preceq_M$ is compatible with $\preceq_{y_2}$, it follows that

$$u(c_1, c_2) = A_{c_2} u_1(c_1) + B_{c_2} = A_{c_1} u_2(c_2) + B_{c_1},$$

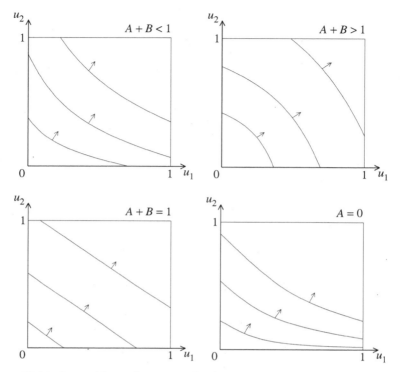

Figure 10.1.   Separable preferences. The three cases considered in section 3.6.2 are illustrated, together with the case $A = 0$ that I suggest should be regarded as the norm. The utility function u is constant along the curves that have been drawn. Economists say these indifference curves are convex in the case $A+B < 1$ and concave in the case $A + B > 1$. In the case $A = 0$, Pandora is indifferent between all values of u_1 with $0 \leqslant u_1 \leqslant 1$ when $u_2 = 0$.

where $B_{c_1} = u_1(c_1)$ and $B_{c_2} = u_2(c_2)$. Thus,

$$u_1(c_1)(A_{c_2} - 1) = u_2(c_2)(A_{c_1} - 1),$$

and so $A_{c_1} = Ju_1(c_1) + 1$ and $A_{c_2} = Ju_2(c_2) + 1$ for some constant J.

It follows that any Von Neumann and Morgenstern utility function u that represents $\preceq$ on lott(C) can be written in the form

$$u(c_1, c_2) = Ju_1(c_1)u_2(c_2) + u_1(c_1) + u_2(c_2) + K.$$

To obtain the representation (3.3), simply renormalize u, u_1, and u_2 as in section 3.6.2. Figure 10.1 illustrates the three cases: $A + B < 1$, $A + B = 1$, and $A + B > 1$. It also shows what happens when $A = 0$, so that an absolute zero can be identified.

10.2 Hausdorff's Paradox of the Sphere

The proof of Hausdorff's paradox is beyond the scope of this book, but it isn't hard to show that the three sets A, B, and C into which he partitions a sphere all have inner measure zero and outer measure one.

If $\underline{m}(A) = a > 0$, then we can find a measurable set $F \subset A$ with $m(F) > \frac{1}{2}a$. The rotations that take A to B and C will take F to G and H, where $m(G) = m(H) = m(F) > \frac{1}{2}a$. Since $G \subset B$ and $H \subset C$, G and H can't overlap, and so $p(G \cup H) = m(G) + m(H) > a$. Hence $\underline{m}(B \cup C) > a$.

But A can be rotated onto $B \cup H$, and so $a = \underline{m}(A) = \underline{m}(B \cup C) > a$. This contradiction implies that $\underline{m}(A) = 0$.

It remains to observe that $\overline{m}(A) = \overline{m}(B \cup C) = 1 - \underline{m}(A) = 1$, because $B \cup C$ is the complement of A in the sphere.

10.3 Conditioning on Zero-Probability Events

The geometric arguments of section 5.5.3 that show how problems can arise when trying to condition on a zero-probability event are illustrated in figure 10.2. Let E denote the event that a randomly chosen chord to a circle of radius r exceeds r. What is $\text{prob}(E)$?

The probability is $\sqrt{3}/2$. We begin with the observation that

$$\text{prob}(E) = \sum_F \text{prob}(E \mid F) \, \text{prob} \, F,$$

where F denotes a member of a finite partition of the sample space B. A limiting case of this result is

$$\text{prob}(E) = \frac{1}{2\pi} \int_0^{2\pi} \text{prob}(E \mid \theta) \, d\theta, \tag{10.1}$$

where $\text{prob}(E \mid \theta)$ is the conditional probability that E is true given that the midpoint of a chord lies on a radius that makes an angle θ with a fixed axis. If the radius of the circle is r, and the distance from the center of the circle to the midpoint of a chord is x, then figure 10.2(a) shows that the length of the chord is $2y = 2\sqrt{r^2 - x^2}$. It follows that $2y < r$ if and only if $x < r\sqrt{3}/2$. If we assume that x is uniformly distributed on the radius, then

$$\text{prob}(E \mid \theta) = \frac{1}{r} \int_0^{r\sqrt{3}/2} dx = \sqrt{3}/2.$$

Substituting this result in (10.1), we find that $\text{prob}(E) = \sqrt{3}/2$.

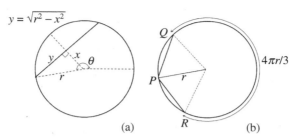

Figure 10.2. How long is a random chord? The left-hand diagram shows that a chord is longer than a radius if and only if $2\sqrt{r^2 - x^2} > r \iff x < r\sqrt{3}/2$. The right-hand diagram shows that a chord with one endpoint at P is longer than a radius if the other endpoint lies between Q and R. The longer arc joining Q and R has length $4\pi r/3$.

The probability is $\frac{3}{4}$. The alternative calculation is based on the formula

$$\text{prob}(E) = \frac{1}{\pi r^2} \int_0^1 \text{prob}(E\,|\,x) 2\pi x\, dx = \frac{1}{\pi r^2} \int_0^{r\sqrt{3}/2} 2\pi x\, dx = \frac{3}{4},$$

where $\text{prob}(E\,|\,x)$ is the conditional probability that E is true, given that the midpoint of a chord lies on a concentric circle of radius x. We therefore have that $\text{prob}(E\,|\,x) = 0$ unless $x < r\sqrt{3}/2$.

The probability is $\frac{2}{3}$. Figure 10.2(b) locates the endpoint of the chord that is chosen first at the point P. We then seek the value of $\text{prob}(E\,|\,P)$.

If the other endpoint of the chord is placed at Q or R, then its length is equal to the radius. The arc of the circle between Q and R has length $4\pi r/3$. Dividing by $2\pi r$, we find that $\text{prob}(E\,|\,P) = \frac{2}{3}$.

Which answer is right? There isn't a one-and-only correct answer. In all three cases, we went beyond what the theory allows by assigning a value to a probability conditioned on a zero-probability event. We implicitly do this whenever we employ calculus in a probabilistic calculation, but *how* we do it should be understood as being part of our underlying model.

The first of the three cases best illustrates that it can matter how we proceed when following Kolmogorov's advice to approximate the zero-probability event by events of positive probability. We can approximate the event of being on a particular radius in two ways, as illustrated in figure 10.3. We implicitly chose the first of these when we assumed that x was uniformly distributed on the radius, and so $\text{prob}(E\,|\,\theta) = \sqrt{3}/2$. If we had chosen the second possibility, then

$$\text{prob}(E\,|\,\theta) = \frac{2}{r^2} \int_0^{r\sqrt{3}/2} x\, dx = \frac{3}{4}.$$

Figure 10.3. Kolmogorov's advice. Kolmogorov recommends considering a sequence of events F_n with $\text{prob}(F_n) > 0$ that converges on a zero-probability event F. The figure shows two possible choices for events F_n that approximate a radius. The probability that the length of a random chord exceeds the length of a radius depends on the choice made.

10.4 Applying the Hahn–Banach Theorem

Section 6.4.3 claims that if Y is the vector space of almost convergent sequences and z lies outside Y, then p can be extended as a Banach limit from Y to z so as to make $p(z)$ equal to any point w we like in the interval $I = [\underline{p}(z), \overline{p}(z)]$. To check that this is true, we appeal to the proof of the Hahn–Banach theorem with $\lambda = \overline{p}$. The reason that we need the extension of p to satisfy $p(x) \leqslant \overline{p}(x)$ for all x is to ensure that the value of $p(x)$ remains unchanged when we throw away a finite number of the terms of x. We can't make do with less than this requirement, because of (6.12).

Recall that the extension is engineered one step at a time (section 6.4.2). When we extend p from Y to z by setting $p(z) = w$, we also need to extend p to the vector space Z spanned by Y and z. Any x in Z can be expressed uniquely in the form $y + \alpha z$. Since a Banach limit is linear, our extension will need to have the property that $p(y + \alpha z) = p(y) + \alpha p(z)$ for all α and all y. The question is then whether $p(y) + \alpha w \leqslant \overline{p}(y + \alpha z)$ for all α and all y in Y.

First observe that, for all y in Y,

$$\overline{p}(y + z) = p(y) + \overline{p}(z),$$

as $\overline{p}(y + z) \leqslant \overline{p}(y) + \overline{p}(z) = p(y) + \overline{p}(z)$, and $\overline{p}(z) = \overline{p}(-y + z + y) \leqslant \overline{p}(-y) + \overline{p}(z + y) = p(-y) + \overline{p}(z + y) = -p(y) + \overline{p}(z + y)$.

When $\alpha \geqslant 0$, we then have that

$$\overline{p}(y + \alpha z) = p(y) + \overline{p}(\alpha z) = p(y) + \alpha \overline{p}(z) \geqslant p(y) + \alpha w;$$

and when $\alpha \leqslant 0$,

$$\overline{p}(y + \alpha z) = p(y) + \overline{p}(\alpha z) = p(y) + \alpha \underline{p}(z) \geqslant p(y) + \alpha w.$$

10.5 Muddling Boxes

Section 6.5 asserts that two different definitions of the upper and lower probability of a muddling box are the same. We consider only upper probabilities here, since the argument for lower probabilities is much the same. The proof we use is adapted from Goffman and Pedrick (1965, p. 68).

The first definition appears in section 6.4.3. The upper probability $\overline{p}(x)$ of a sequence x whose terms lie in the interval $[0, 1]$ is defined to be the smallest real number $\overline{p}$ for which it is true that for any $\epsilon > 0$, there exists N such that for any $n > N$ and all values of m,

$$\frac{1}{n} \sum_{j=1}^{n} x_{m+j} < \overline{p} + \epsilon. \tag{10.2}$$

The second definition appears in section 6.5. The upper probability $\overline{\pi}(x)$ of the sequence x is defined as

$$\overline{\pi} = \inf \left\{ \limsup_{n \to \infty} \frac{1}{k} \sum_{\ell=1}^{k} x_{m_\ell + n} \right\}, \tag{10.3}$$

where the infimum ranges over all finite sets $\{m_1, m_2, \ldots, m_k\}$ of natural numbers.

Proof that $\overline{\pi} \leqslant \overline{p}$. Suppose that $\epsilon > 0$. For a large enough value of k and all values of m,

$$\frac{1}{k} \sum_{\ell=1}^{k} x_{m+\ell} < \overline{p} + \epsilon.$$

Take $m_1 = 1$, $m_2 = 2$, $\ldots$, $m_k = k$ in (10.3). Then

$$\overline{\pi} \leqslant \limsup_{n \to \infty} \frac{1}{k} \sum_{\ell=1}^{k} x_{m_\ell + n} \leqslant \overline{p} + \epsilon.$$

Since $\overline{\pi} \leqslant \overline{p} + \epsilon$ for all $\epsilon > 0$, it follows that $\overline{\pi} \leqslant \overline{p}$.

Proof that $\overline{p} \leqslant \overline{\pi}$. Suppose that $\epsilon > 0$. We can find values of k, m_1, m_2, $\ldots$, m_k such that

$$\limsup_{n \to \infty} \frac{1}{k} \sum_{\ell=1}^{k} x_{m_\ell + n} < \overline{\pi} + \epsilon.$$

Hence, there exists N such that for all $n \geqslant N$,

$$\frac{1}{k} \sum_{\ell=1}^{k} x_{m_\ell + n} < \overline{\pi} + \epsilon. \tag{10.4}$$

$$x_6 - x_3, \; x_7 - x_4, \; x_8 - x_5, \; x_9 - x_6, \; x_{10} - x_7,$$
$$x_{11} - x_8, \; x_{12} - x_9, \; x_{13} - x_{10}, \; x_8 - x_3, \; x_9 - x_4, \; x_{10} - x_5,$$
$$x_{11} - x_6, \; x_{12} - x_7, \; x_{13} - x_8, \; x_{14} - x_9, \; x_{15} - x_{10}$$

Figure 10.4. Cancelations. The figure expands the double summation that appears as the final term of (10.5) for the case $n = 2$, $k = 2$, $m_1 = 3$, $m_2 = 5$, $i = 8$. Only $6 + 10 = 2(m_1 + m_2)$ terms fail to cancel.

Replace n by $n - N$ and m_ℓ by $m_\ell + N$, so that (10.4) then holds for all values of n. Write $n + j$ for n in the left-hand side of the new inequality and average the result to get

$$\frac{1}{i} \sum_{j=1}^{i} \frac{1}{k} \sum_{\ell=1}^{k} x_{m_\ell + n + j} = \frac{1}{ik} \sum_{\ell=1}^{k} \sum_{j=1}^{i} \{x_{j+n} + (x_{m_\ell + n + j} - x_{j+n})\}$$

$$= \frac{1}{i} \sum_{j=1}^{i} x_{j+n} + \frac{1}{ik} \sum_{\ell=1}^{k} \sum_{j=1}^{i} (x_{m_\ell + n + j} - x_{j+n}). \quad (10.5)$$

Figure 10.4 illustrates why cancelations in the concluding double summation leave only at most $2(m_1 + m_2 + \cdots + m_k)$ surviving terms. It follows that the double summation can be made smaller in modulus than ϵ by taking i sufficiently large.

It remains to observe that if i is sufficiently large,

$$\frac{1}{i} \sum_{j=1}^{i} x_{j+n} < \overline{\pi} + 2\epsilon.$$

Hence, for every $\epsilon > 0$, $\overline{p} \leqslant \overline{\pi} + 2\epsilon$, and so $\overline{p} \leqslant \overline{\pi}$.

10.6 Solving a Functional Equation

Solving the functional equation $f(xy) = f(x)f(y)$ for continuously differentiable real functions is a standard exercise in calculus. We first differentiate the equation partially with respect to y to obtain that $xf'(xy) = f(x)f'(y)$. Writing $y = 1$ in this equation, we are led to the differential equation

$$\frac{f'(x)}{f(x)} = \frac{\alpha}{x},$$

where $\alpha = f'(1)$. We need the derivative of f to be continuous in order to integrate this equation to obtain that

$$f(x) = Kx^\alpha,$$

for some constant K. Substituting this formula back into the functional equation with which we began, we find that $K = 0$ or $K = 1$.

Applying the same technique to the functional equation $U(pq, PQ) = U(p, P)U(q, Q)$, we find that

$$U(p, P) = K(P)p^\alpha \quad \text{and} \quad U(p, P) = L(p)P^\beta,$$

where $\alpha = U_1(1, 1)$ and $\beta = U_2(1, 1)$. Thus

$$\frac{K(P)}{P^\beta} = \frac{L(p)}{p^\alpha} = c,$$

where c is an absolute constant.

Since $U(p, p) = p$ implies that $c = 1$ and $\alpha + \beta = 1$, we find that the solution to our functional equation takes the form

$$U(p, P) = p^{1-h}P^h,$$

where the Hurwicz coefficient h satisfies $0 \leqslant h \leqslant 1$.

10.7 Additivity

Section 9.2.2 considers the behavior of $\pi(E) + \pi(\sim E)$. For the case when $h < \frac{1}{2}$ and $0 < p < P < 1$, we have that

$$f(p, P) = p^{1-h}P^h + (1 - P)^{1-h}(1 - p)^h$$
$$< p^{1/2}P^{1/2} + (1 - P)^{1/2}(1 - p)^{1/2}$$
$$< \tfrac{1}{2}(p + P + 1 - P + 1 - p) = 1.$$

To justify the first step, differentiate the left-hand side with respect to h to confirm that it is a strictly increasing function of h in the relevant range. To justify the second step, apply the inequality of the arithmetic and geometric means.

To confirm that there are cases with $h > \frac{1}{2}$ for which $\pi(E) + \pi(\sim E) > 1$, check that $f_1(\frac{1}{2}, \frac{1}{2}) = 1 - 2h < 0$ for $h > \frac{1}{2}$. Hence, $f(p, \frac{1}{2}) > f(\frac{1}{2}, \frac{1}{2}) = 1$ for some values of $p < \frac{1}{2}$.

10.8 Muddled Equilibria in Game Theory

Section 9.3.1 describes some new Nash equilibria that arise in the Battle of the Sexes when muddled strategies are allowed. This section justifies these results.

We work with the more general form of the Battle of the Sexes given in figure 10.5. The payoffs in this version of the game are to be interpreted

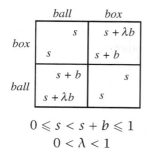

$$0 \leqslant s < s + b \leqslant 1$$
$$0 < \lambda < 1$$

For the mixed equilibrium, play your second pure strategy with probability

$$e = \frac{\lambda}{1 + \lambda}.$$

To secure your maximum payoff, play your second pure strategy with probability

$$m = \frac{1}{1 + \lambda}.$$

Figure 10.5. A general version of the Battle of the Sexes. The payoffs are to be interpreted as probabilities of winning in win-or-lose lotteries in which losing corresponds to absolute zero on the player's utility scale. The Battle of the Sexes of figure 9.5 is a scaled-up version of the case with $s = 0$, $b = 1$, and $\lambda = \frac{1}{2}$. The text also analyzes the case $s = b = \frac{1}{2}$ and $\lambda = \frac{1}{2}$.

as the probabilities of players winning win-or-lose lotteries in which losing corresponds to an absolute zero on their utility scales. Notice that the value of s is strategically significant when muddled strategies are used (although each of a player's payoffs can be multiplied by the same positive factor without altering any strategic considerations).

The standard version of the Battle of the Sexes given in figure 9.5 corresponds to the case when $s = 0$ and $\lambda = \frac{1}{2}$. The case when $s = 0$ and $\lambda = 1$ corresponds to the Driving Game. However, these cases are rather drastic since a failure to coordinate results in both players receiving their absolute zero. For everyday problems, we need to allow s to be positive (as in section 9.2.3).

The analysis of the Battle of the Sexes given in section 9.3 generalizes to the version given in figure 10.5. There are two asymmetric pure Nash equilibria and a symmetric mixed Nash equilibrium. In the latter, both players use their second pure strategy with probability $e = \lambda/(1 + \lambda)$. Their expected payoff is then $c = s + eb$, which is also their maximin value in the game. But the strategy that guarantees the maximin value requires that the second pure strategy be played with probability $m = 1/(1 + \lambda)$. Since $m \neq e$ when $\lambda \neq 1$, the general version of the Battle of the Sexes generates the same problem in identifying a solution to the Battle of the Sexes in a symmetric environment that we encountered in section 9.3.

If Adam plays his second pure strategy with probability p and Eve plays her second pure strategy with probability q, then Adam's payoff is

$$\pi_1(p,q) = c - b(1 + \lambda)(p - m)(q - e),$$

where $c = s + b\lambda/(1 + \lambda)$ is the mixed equilibrium payoff (that each player gets when $p = q = e$). It is also the maximin payoff each player guarantees by playing $p = m$ or $q = m$.

Plan of attack. The first item that needs to be proved is that the Battle of the Sexes has a continuum of symmetric Nash equilibria when muddled strategies are allowed. We assume that both Adam and Eve have the same Hurwicz coefficient h, which satisfies $0 < h \leqslant \frac{1}{2}$. (It is easy to get results if $h > \frac{1}{2}$, but not very interesting.) We write $\alpha = 1 - h$ and $\beta = h$.

Suppose that Eve uses a muddled strategy with upper probability $\overline{q}$ and lower probability $\underline{q}$. In a symmetric equilibrium, Adam's best reply will be the muddled strategy in which $\overline{p} = \overline{q}$ and $\underline{p} = \underline{q}$.

We will only look at muddled strategies for Eve for which Adam's best reply in *mixed* strategies is $p = \underline{q}$. The next step is then to show that Adam gets the same payoff if he deviates from this mixed strategy to the muddled strategy in which $\underline{p} = p$ and $\overline{p} = \overline{q}$. This will require further restrictions on Eve's muddled strategy, but we won't worry about this second step until we have finished studying the question of Adam's best reply in mixed strategies.

Adam's best reply in mixed strategies. Our first restriction on Eve's choice of muddled strategy is that $\underline{q} < e < \overline{q}$. If we write $y = \underline{q} - e < 0$ and $Y = \overline{q} - e > 0$, then Adam's best reply in mixed strategies is found by locating the value of $x = p - m$ in the range $[-m, 1 - m]$ that maximizes

$$g = \begin{cases} b(1 + \lambda)(A - xy)^{\alpha}(A - xY)^{\beta}, & \text{if } x \leqslant 0, \\ b(1 + \lambda)(A - xY)^{\alpha}(A - xy)^{\beta}, & \text{if } x \geqslant 0, \end{cases}$$

where $A = c/b(1 + \lambda)$.

The plan is to fix $y < 0$ and $Y > 0$ so that the solution to Adam's maximization problem is

$$x = y + e - m. \tag{10.6}$$

It will then follow that $p = x + m = y + e = \underline{q}$.

Such a solution x to our maximization problem will necessarily satisfy $x < 0$ (because $y < 0$ and $e < m$). If x doesn't occur at an endpoint of the admissible interval $[-m, 1 - m]$, it can be found by differentiating $\{A - xy\}^{\alpha}\{A - xY\}^{\beta}$ with respect to x and setting the result equal to zero. The resulting x is given by

$$x = A\left\{\frac{\alpha}{Y} + \frac{\beta}{y}\right\}. \tag{10.7}$$

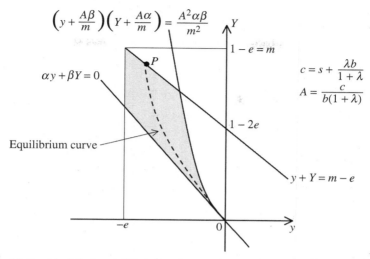

Figure 10.6. Muddled equilibria in the Battle of the Sexes. All points (y, Y) on the dashed curve correspond to symmetric Nash equilibria in muddled strategies. The point $(0, 0)$ corresponds to the mixed Nash equilibrium. As we move (y, Y) away from $(0, 0)$ along the dashed curve, the payoff to each player declines at first. However, there are parameter values for which the payoff to the players at the point P where the dashed curve crosses the boundary of the shaded set exceeds the payoff they get at the mixed equilibrium.

To make this conclusion compatible with our assumptions, we need to insist that

$$-m \leqslant A \left\{ \frac{\alpha}{Y} + \frac{\beta}{y} \right\} < 0,$$

which is equivalent to the requirements that $\alpha y + \beta Y > 0$ and

$$\left(y + \frac{A\beta}{m} \right) \left(Y + \frac{A\alpha}{m} \right) \leqslant \frac{A^2 \alpha \beta}{m^2}. \tag{10.8}$$

These are two of the constraints on the pair (y, Y), as illustrated in figure 10.6. The constraints $-e \leqslant y < 0$ and $0 < Y \leqslant 1 - e$ simply recognize that probabilities lie in $[0, 1]$. The constraint $y + Y \leqslant m - e$ is explained later by inequality (10.10).

Our restrictions on y and Y justify the claim that the value of x given by (10.7) maximizes Adam's payoff for values of x in the interval $[-m, 0]$, but what of the interval $[0, 1 - m]$? To show that (10.7) is also maximal when this interval is admitted, we need to use the fact that $\alpha \geqslant \beta$ (because $h \leqslant \frac{1}{2}$). This inequality implies that $(A - xY)^\alpha (A - xy)^\beta$ decreases on $[0, 1 - m]$, and hence is maximized at $x = 0$, where it is equal to $(A - xy)^\alpha (A - xY)^\beta$.

Equilibrium condition. To make Adam's best reply $p = \underline{q}$, we appeal to (10.7) and set

$$y + e - m = A\left\{\frac{\alpha}{Y} + \frac{\beta}{y}\right\},$$

from which we obtain the equation

$$Y = \frac{A\alpha y}{y^2 + y(e - m) - A\beta}. \tag{10.9}$$

This equation defines the dashed curve in figure 10.6. The curve has a vertical asymptote $y = -\eta$ for some positive value of η. On the interval $(-\eta, 0]$, it lies above the line defined by $\alpha y + \beta Y = 0$ and beneath the branch of the hyperbola

$$\left(y + \frac{A\beta}{m}\right)\left(Y + \frac{A\alpha}{m}\right) = \frac{A^2 \alpha \beta}{m^2}$$

drawn in figure 10.6.

Adam's best reply in muddled strategies. If Eve chooses the muddled strategy corresponding to the pair (y, Y), we have found conditions under which it is a best reply for Adam to use the mixed strategy $(x, X) = (y + e - m, y + e - m)$. We now show that it remains a best reply for Adam to use the muddled strategy $(x, X) = (y + e - m, Y + e - m)$. The proof consists of showing that the deviation doesn't affect Adam's best or worst outcome, given that Eve has chosen (y, Y). For this purpose, we need the following inequalities:

$$A - (y + e - m)y \leqslant A - (Y + e - m)y \leqslant A - (y + e - m)Y,$$

$$A - (y + e - m)y \leqslant A - (Y + e - m)Y \leqslant A - (y + e - m)Y.$$

Only the inequality $A - (y + e - m)y \leqslant A - (Y + e - m)Y$ fails to be satisfied whenever $y < Y$. To satisfy this exceptional inequality when $y < Y$, we need an extra condition:

$$y + Y \leqslant m - e. \tag{10.10}$$

This is the constraint illustrated in figure 10.6 that was left unexplained.

A continuum of muddled equilibria. Every pair (y, Y) on the dashed line in figure 10.6 that also lies to the right of the line $y = -e$ and beneath the line $y + Y = m - e$ corresponds to a symmetric Nash equilibrium in muddled strategies. A continuum of such equilibria therefore exists.

The case $s = 0$, $b = 1$, $\lambda = h = \frac{1}{2}$. This is the case considered in figure 9.5, with the extra assumption that the players are uncertainty neutral ($h = \frac{1}{2}$). The value of b only scales the payoffs up or down.

In this case, the payoff to each player at the point P drawn in figure 10.6 is approximately 0.78. Each player uses a muddled strategy with lower probability approximately $\frac{1}{6}$ and upper probability approximately $\frac{5}{6}$.

The payoff pair $(0.78, 0.78)$ is indicated in figure 9.7 by a star. It lies outside the mixed-strategy noncooperative payoff region of the Battle of the Sexes, because the highest payoff the players can get if both play the same mixed strategy is 0.75.

By continuity, similar results can be obtained for values of $h < \frac{1}{2}$, but one can't reduce h very much before only symmetric equilibria with worse payoffs than the mixed equilibrium are left.

The case $s = b = \lambda = h = \frac{1}{2}$. In this everyday case, the point P in figure 10.6 moves to the corner of the shaded region. At this equilibrium each player uses a muddled strategy with lower probability zero and upper probability one. The payoff to each player at the equilibrium is $\sqrt{2}/2$, which exceeds the payoff of $\frac{2}{3}$ that each player gets at the mixed equilibrium.

The Driving Game. Taking the limit as $\lambda \to 1$ generates the Driving Game of figure 9.5. It is mentioned only to confirm that no new equilibria are created by allowing muddled strategies.

References

Anscombe, F., and R. Aumann. 1963. A definition of subjective probability. *Annals of Mathematical Statistics* 34:199–205.

Arrow, K. 1959. Rational choice functions and ordering. *Economica* 26:121–27.

———. 1963. *Social Choice and Individual Values*. New Haven, CT: Yale University Press.

———. 1971. *Essays on the Theory of Risk Bearing*. Chicago, IL: Markham.

Aumann, R. 1995. Backward induction and common knowledge of rationality. *Games and Economic Behavior* 8:6–19.

Aumann, R., and M. Maschler. 1972. Some thoughts on the minimax principle. *Management Science* 18:P54–P63.

Barbara, S., and M. Jackson. 1988. Maximin, leximin, and the protective criterion: characterizations and comparisons. *Journal of Economic Theory* 46:34–44.

Bentham, J. 1863. Pannonial fragments. In *Works of Jeremy Bentham III* (ed. J. Bowring). Edinburgh: W. Tait.

Berger, J. 1985. *Statistical Decision Theory and Bayesian Analysis*, Springer Series in Statistics. Berlin: Springer.

Bertrand, J. 1889. *Calcul des Probabilités*. Paris: Gauthier-Villars.

Billingsley, C. P. 1986. *Probability and Measure*, 2nd edn. New York: Wiley.

Bingham, N. 2005. Finite additivity versus countable additivity: on the centenary of Bruno de Finetti. Working Paper, Mathematics Department, Imperial College London.

Binmore, K. 1984. Bargaining conventions. *International Journal of Game Theory* 13:193–200.

———. 1994. *Playing Fair: Game Theory and the Social Contract I*. Cambridge, MA: MIT Press.

———. 1996. A note on backward induction. *Games and Economic Behavior* 17: 135–37.

———. 1997. Rationality and backward induction. *Journal of Economic Methodology* 4:23–41.

———. 1998. *Just Playing: Game Theory and the Social Contract II*. Cambridge, MA: MIT Press.

———. 2005. *Natural Justice*. New York: Oxford University Press.

———. 2007a. *Does Game Theory Work? The Bargaining Challenge*. Cambridge, MA: MIT Press.

———. 2007b. *Playing for Real*. New York: Oxford University Press.

Binmore, K., and A. Voorhoeve. 2003. Defending transitivity against Zeno's paradox. *Philosophy and Public Affairs* 31:272–79.

———. 2006. Transitivity, the sorites paradox, and similarity-based decision-making. *Erkenntnis* 64:101–14.

Blackburn, S. 1998. *Ruling Passions*. Oxford: Oxford University Press.

Borel, É. 1905. Sur les probabilités dénombrables et leurs applications arithmétiques. *Rendiconti Del Circolo Matematico Di Palermo* 29:247-71.

Brams, S. 1975. Newcomb's problem and Prisoners' Dilemma. *Journal of Conflict Resolution* 19:5612.

Breese, J., and K. Fertig. 1991. Decision making with interval influence diagrams. In *Uncertainty in Artificial Intelligence* (ed. P. Bonissone, M. Henrion, N. Kanal, and J. Lenner). Dordrecht: Elsevier.

Broome, J. 1991. *Weighing Goods*. Oxford: Blackwell.

Calude, C. 1998. *Information and Randomness*. Berlin: Springer.

Camerer, C., and D. Harless. 1994. The predictive utility of generalized expected utility theories. *Econometrica* 62:1251-90.

Carnap, R. 1937. *The Logical Syntax of Language*. London: Kegan Paul.

——. 1950. *Logical Foundations of Probability*. Chicago, IL: University of Chicago Press.

Chaitin, G. 2001. *Exploring Randomness*. Berlin: Springer.

Chernov, H. 1954. Rational selection of decision functions. *Econometrica* 22: 422-43.

Chrisman, L. 1995. Incremental conditioning of upper and lower probabilities. *International Journal of Approximate Reasoning* 13:1-25.

Church, A. 1940. On the concept of a random sequence. *Bulletin of the American Mathematical Society* 46:130-35.

Clark, M. 2002. *Paradoxes from A to Z*. London: Routledge.

Damasio, A. 1994. *Descartes' Error: Emotion, Reason and the Human Brain*. New York: Avon Books.

Dawid, A. P., and V. Vovk. 1999. Prequential probability: principles and properties. *Bernoulli* 5:125-62.

de Finetti, B. 1937. La prévision: ses lois logique, ses sources subjectives. *Annales de l'Institut Henri Poincaré* 7:1-68.

——. 1974a. *Theory of Probability*, volume I. New York: Wiley.

——. 1974b. *Theory of Probability*, volume II. New York: Wiley.

DeGroot, M. 1978. *Optimal Statistical Decisions*. New York: McGraw-Hill.

Dempster, A. 1967. Upper and lower probabilities induced by a multivalued mapping. *Annals of Mathematical Statistics* 38:325-39.

Diaconis, P., and D. Freedman. 1986. On the consistency of Bayes estimates. *Annals of Statistics* 14:1-26.

Dow, J., and S. Werlang. 1992. Uncertainty aversion, risk aversion, and the optimal choice of portfolio. *Econometrica* 60:197-204.

——. 1994. Nash equilibrium under Knightian uncertainty: breaking down backward induction. *Journal of Economic Theory* 64:302-24.

Earman, J. 1992. *Bayes or Bust?* Cambridge, MA: MIT Press.

Ellsberg, D. 1961. Risk, ambiguity, and the Savage axioms. *Quarterly Journal of Economics* 75:643-69.

Elster, J. 1985. *Making Sense of Marx*. Cambridge: Cambridge University Press.

Elster, J., and J. Roemer. 1992. *Interpersonal Comparisons of Well-Being*. Cambridge: Cambridge University Press.

Epstein, L. 1999. A definition of uncertainty aversion. *Review of Economic Studies* 66:579-608.

——. 2001. Sharing ambiguity. *American Economic Review* 91:45-50.

Fertig, K., and J. Breese. 1990. Interval influence diagrams. In *Uncertainty in Artificial Intelligence* (ed. M. Henrion, P. Shachter, L. Kanal, and J. Lenner). Dordrecht: Elsevier.

Fine, T. 1973. *Theories of Probability.* London: Academic Press.

——. 1988. Lower probability models for uncertainty and nondeterministic processes. *Journal of Statistical Planning and Inference* 20:389–441.

Fishburn, P. 1970. *Utility Theory for Decision Making.* New York: Wiley.

——. 1982. *Foundations of Expected Utility.* Amsterdam: Kluwer.

Fishburn, P., and A. Rubinstein. 1982. Time preference. *International Economic Review* 23:677–94.

Friedman, M., and L. Savage. 1948. Utility analysis of choices involving risk. *Journal of Political Economy* 56:179–304.

Gardenfors, P., and N. Sahlin. 1982. Unreliable probabilities, risk taking and decision making. *Synthese* 53:361–86.

Gauthier, D. 1993. Uniting separate persons. In *Rationality, Justice and the Social Contract* (ed. D. Gauthier and R. Sugden). Hemel Hempstead, UK: Harvester Wheatsheaf.

Ghirardato, P., and M. Marinacci. 2002. Ambiguity made precise: a comparative foundation. *Journal of Economic Theory* 102:251–89.

Ghirardato, P., F. Maccheroni, and M. Marinacci. 2004. Differentiating ambiguity and ambiguity attitude. *Journal of Economic Theory* 118:133–73.

Gibbard, A. 1990. *Wise Choices and Apt Feelings: A Theory of Normative Judgment.* Oxford: Clarendon Press.

Gigerenzer, G. 1996. On narrow norms and vague heuristics: a reply to Kahneman and Tversky. *Psychological Review* 103:582–91.

Gilboa, I. 2004. *Uncertainty in Economic Theory: Essays in Honor of David Schmeidler's 65th Birthday.* London: Routledge.

Gilboa, I., and D. Schmeidler. 2001. *A Theory of Case-Based Decisions.* Cambridge: Cambridge University Press.

——. 2004. Maximin expected utility with non-unique priors. In *Uncertainty in Economic Theory: Essays in Honor of David Schmeidler's 65th Birthday* (ed. I. Gilboa). London: Routledge.

Gillies, D. 2000. *Philosophical Theories of Probability.* London: Routledge.

Giron, F., and S. Rios. 1980. Quasi-Bayesian behavior; a more realistic approach to decision-making? In *Bayesian Statistics* (ed. J. Bernado, J. DeGroot, D. Lindley, and A. Smith). Valencia, Spain: Valencia University Press.

Goffman, C., and G. Pedrick. 1965. *First Course in Functional Analysis.* Englewood Ciffs, NJ: Prentice-Hall.

Good, I. J. 1983. *Good Thinking: The Foundations of Probability and Its Applications.* Minneapolis: University of Minnesota Press.

Goodin, R. 1995. *Utilitarianism as a Public Philosophy.* Cambridge: Cambridge University Press.

Gorman, W., and G. Myles. 1988. Characteristics. In *The New Palgrave: A Dictionary of Economics* (ed. J. Eatwell, M. Milgate, and P. Newman). London: Macmillan.

Greenberg, J. 2000. The right to remain silent. *Theory and Decision* 48:193–204.

Hacking, I. 1975. *The Emergence of Probability. A Philosophical Study of Early Ideas about Probability, Induction and Statistical Inference.* Cambridge: Cambridge University Press.

Halpern, J., and R. Fagin. 1992. Two views of belief: belief as generalized probability and belief as evidence. *Artificial Intelligence* 54:275–317.

Hammond, P. 1988. Consequentialist foundations for expected utility. *Theory and Decision* 25:25–78.

——. 1992. Harsanyi's utilitarian theorem: a simpler proof and some ethical connotations. In *Rational Interaction: Essays in Honor of John Harsanyi* (ed. R. Selten). Berlin: Springer.

——. 1999. Uncertainty. In *The New Palgrave: A Dictionary of Economics* (ed. J. Eatwell, M. Milgate, and P. Newman). London: Macmillan.

Hardin, R. 1988. *Morality within the Limits of Reason.* Chicago, IL: University of Chicago Press.

Hardy, G. H. 1956. *Divergent Series.* Oxford: Oxford University Press.

Harsanyi, J. 1953. Cardinal utility in welfare economics and in the theory of risk-taking. *Journal of Political Economy* 61:434–35.

——. 1955. Cardinal welfare, individualistic ethics, and the interpersonal comparison of utility. *Journal of Political Economy* 63:309–21.

——. 1964. A general solution for finite non-cooperative games, based on risk dominance. In *Advances in Game Theory* (ed. L. Shapley, M. Dresher, and A. Tucker). Princeton, NJ: Princeton University Press.

——. 1966. A general theory of rational behavior in game situations. *Econometrica* 34:613–34.

——. 1977. *Rational Behavior and Bargaining Equilibrium in Games and Social Situations.* Cambridge: Cambridge University Press.

Henrich, J., R. Boyd, S. Bowles, C. Camerer, E. Fehr, and H. Gintis. 2004. *Foundations of Human Sociality: Economic Experiments and Ethnographic Evidence from Fifteen Small-Scale Societies.* New York: Oxford University Press.

Hey, J., and C. Orme. 1994. Investigating generalizations of expected utility theory using experimental data. *Econometrica* 62:1251–90.

Hintikka, J. 1962. *Knowledge and Belief: An Introduction to the Logic of the Two Notions.* Ithaca, NY: Cornell University Press.

Hirshleifer, J., and J. Riley. 1992. *The Analytics of Uncertainty and Information.* Cambridge: Cambridge University Press.

Hobbes, T. 1986. *Leviathan* (ed. C. B. Macpherson). London: Penguin Classics. (First published in 1651.)

Howson, C. 2000. *Hume's Problem.* Oxford: Oxford University Press.

Huber, P. 1980. *Robust Statistics.* New York: Wiley.

Hume, D. 1975. *Enquiries Concerning Human Understanding and Concerning the Principles of Morals* (ed. L. A. Selby-Bigge; revised by P. Nidditch), 3rd edn. Oxford: Clarendon Press. (First published in 1777.)

——. 1978. *A Treatise of Human Nature* (ed. L. A. Selby-Bigge; revised by P. Nidditch), 2nd edn. Oxford: Clarendon Press. (First published in 1739.)

Hurwicz, L. 1951. Optimality criteria for decision making under ignorance. Cowles Statistics Paper 370.

Jaynes, E., and L. Bretthorst. 2003. *Probability Theory: The Logic of Science.* Cambridge: Cambridge University Press.

Jeffrey, R. 1965. *The Logic of Decision.* New York: McGraw-Hill.

Jevons, S. 1871. *The Theory of Political Economy.* London: Macmillan.

Kadane, J., and A. O'Hagan. 1995. Using finitely additive probability: uniform distributions on the natural numbers. *Journal of the American Statistical Association* 90:626–31.

Kadane, J., M. Schervish, and T. Seidenfeld. 1999. *Rethinking the Foundations of Statistics.* Cambridge: Cambridge University Press.

Kagel, J., R. Battalio, and L. Green. 1995. *Economic Choice Theory: An Experimental Analysis of Animal Behavior.* Cambridge: Cambridge University Press.

Kahneman, D., and A. Tversky. 1973. On the psychology of prediction. *Psychological Review* 80:237–51.

———. 1979. Prospect theory: an analysis of decision under risk. *Econometrica* 47:263–91.

Kant, I. 1998. *Critique of Pure Reason* (translated and intoduced by A. Guyer and A. Wood). Cambridge: Cambridge University Press. (First published in 1781.)

Kaplan, D., and R. Montague. 1960. A paradox regained. *Notre Dame Journal of Formal Logic* 1:79–90.

Karni, E. 1985. *Decision Making under Uncertainty.* Cambridge, MA: Harvard University Press.

Keeney, R., and H. Raiffa. 1975. Additive value functions. In *Théorie de la Décision et Applications* (ed. J. Ponssard et al.). Paris: Fondation Nationale pour l'Enseignement de la Gestion des Entreprises.

Kelsey, D. 1992. Theories of choice under ignorance and uncertainty. *Journal of Economic Surveys* 6:133–35.

Kelsey, D., and J. Eichberger. 2000. Non-additive beliefs and strategic equilibrium. *Games and Economic Behavior* 30:83–215.

Keynes, J. M. 1921. *A Treatise on Probability.* London: Macmillan.

Klibanoff, P., and E. Hanany. 2007. Updating preferences with multiple priors. *Theoretical Economics* 2:281–98.

Klibanoff, P., M. Marinacci, and S. Mukerji. 2005. A smooth model of decision-making under ambiguity. *Econometrica* 73:1849–92.

Knight, F. 1921. *Risk, Uncertainty, and Profit.* Boston, MA: Houghton-Mifflin.

Kolmogorov, A. 1950. *Foundations of the Theory of Probability.* New York: Chelsea.

———. 1968. Three approaches for defining the concept of information quantity. *Problems of Information Transmission* 14:3–11.

Kreps, D. 1988. *Notes on the Theory of Choice.* Boulder, CO: Westview Press.

Kuhn, H. 1953. Extensive games and the problem of information. In *Contributions to the Theory of Games II* (ed. H. Kuhn and A. Tucker). Princeton, NJ: Princeton University Press.

Kyburg, H. 1987. Bayesian and non-Bayesian evidential updating. *Artifical Intelligence* 31:271–93.

Layard, R. 2005. *Happiness: Lessons from a New Science.* London: Allen Lane.

Ledyard, J. 1995. Public goods: a survey of experimental research. In *Handbook of Experimental Economics* (ed. J. Kagel and A. Roth). Princeton, NJ: Princeton University Press.

Levi, I. 1980. *The Enterprise of Knowledge.* Cambridge, MA: MIT Press.

———. 1986. *Hard Choices: Decision Making under Unresolved Conflict.* Cambridge: Cambridge University Press.

Lewis, D. 1976. *Counterfactuals.* Oxford: Blackwell.

Lewis, D. 1979. Prisoners' Dilemma as a Newcomb problem. *Philosophy and Public Affairs* 8:235–40.

Lindley, D. 1988. Thomas Bayes (1702–1761). In *The New Palgrave: A Dictionary of Economics* (ed. J. Eatwell, M. Milgate, and P. Newman). London: Macmillan.

Lo, C. 1996. Equilibrium in beliefs under uncertainty. *Journal of Economic Theory* 71:443–84.

———. 1999. Extensive form games with uncertainty averse players. *Games and Economic Behavior* 28:256–70.

Lott, J. 1997. *Uncertainty and Economic Evolution: Essays in Honour of Armen Alchian.* London: Taylor and Francis.

Luce, R., and H. Raiffa. 1957. *Games and Decisions.* New York: Wiley.

Machina, M. 2004. Nonexpected utility theory. In *Encyclopedia of Actuarial Science* (ed. J. Teugels and B. Sundt). Chichester, UK: Wiley.

Marinacci, M. 2000. Ambiguous games. *Games and Economic Behavior* 31:191–219.

———. 2002. Probabilistic sophistication and multiple priors. *Econometrica* 70:755–64.

Marinacci, M., and L. Montrucchio. 2004. Introduction to the mathematics of ambiguity. In *Uncertainty in Economic Theory: Essays in Honor of David Schmeidler's 65th Birthday* (ed. I. Gilboa). London: Routledge.

Mas-Collel, A., M. Whinston, and J. Green. 1995. *Microeconomic Theory.* New York: Oxford University Press.

Maxwell, N. 1998. *The Comprehensibility of the Universe: A New Conception of Science.* Oxford: Oxford University Press.

Mill, J. S. 1962. On liberty. In *Utilitarianism* (ed. M. Warnock). London: Collins. (Essay first published in 1859.)

Milnor, J. 1954. Games against Nature. In *Decision Processes* (ed. R. Thrall, C. Coombs, and R. Davies). New York: Wiley.

Montesano, A., and F. Giovannoni. 1991. Uncertainty aversion and aversion to increasing uncertainty. *Theory and Decision* 41:133–48.

Moore, G. E. 1988. *Principia Ethica.* Buffalo, NY: Prometheus Books. (First published in 1902.)

Mukerji, S., and H. Shin. 2002. Equilibrium departures from common knowledge in games with non-additive expected utility. *Advances in Theoretical Economics* 2:1011.

Nash, J. 1950. The bargaining problem. *Econometrica* 18:155–62.

———. 1951. Non-cooperative games. *Annals of Mathematics* 54:286–95.

Nau, R., and K. McCardle. 1990. Coherent behavior in noncooperative games. *Journal of Economic Theory* 50:424–44.

Nozick, R. 1969. Newcomb's problem and two principles of choice. In *Essays in Honor of Carl G. Hempel* (ed. N. Rescher). Dordrecht: Reidel.

Pearl, J. 1988. On probability intervals. *International Journal of Approximate Reasoning* 2:211–16.

Popper, K. 1959. *The Logic of Scientific Discovery.* London: Hutchison.

———. 1963. *Conjectures and Refutations.* London: Routledge and Kegan Paul.

Quine, W. 1976. *The Ways of Paradox and Other Essays.* Cambridge, MA: Harvard University Press.

———. 1990. The puzzle of the self-torturer. *Philosophical Studies* 59:79–90.

Rachels, S. 1998. Counterexamples to the transitivity of better than. *Australasian Journal of Philosophy* 76:71-83.

——. 2001. Intransitivity. In *The Encyclopedia of Ethics* (ed. L. Becker and C. Becker), 2nd edn., volume II. London: Routledge.

Ramsey, F. 1931. Truth and probability. In *Foundations of Mathematics and Other Logical Essays* (ed. F. Ramsey). New York: Harcourt.

Rao, K., and M. Rao. 1983. *Theory of Charges.* Oxford: Academic Press.

Rawls, J. 1972. *A Theory of Justice.* Oxford: Oxford University Press.

Rényi, A. 1977. *Letters on Probabilty.* Detroit, MI: Wayne State University Press.

Richter, M. 1988. Revealed preference theory. In *The New Palgrave: A Dictionary of Economics* (ed. J. Eatwell, M. Milgate, and P. Newman). London: Macmillan.

Robbins, L. 1938. Inter-personal comparisons of utility. *Economic Journal* 48: 635-41.

Ryan, M. 2002. What do uncertainty-averse players believe? *Economic Theory* 20:47-65.

Samuelson, P. 1947. *Foundations of Economic Analysis.* Cambridge, MA: Harvard University Press.

Savage, L. 1951. *The Foundations of Statistics.* New York: Wiley.

Schmeidler, D. 2004. Subjective probability and expected utility without additivity. In *Uncertainty in Economic Theory: Essays in Honor of David Schmeidler's 65th Birthday* (ed. I. Gilboa). London: Routledge.

Schmidt, D., and T. Neugebauer. 2007. Testing expected utility in the presence of errors. *Economic Journal* 117:470-85.

Seidenfeld, T. 1993. Outline of a theory of partially ordered preferences. *Philosophical Topics* 21:173-88.

Sen, A. 1976. Welfare inequalities and Rawlsian axiomatics. *Theory and Decision* 7:243-62.

——. 1992. *Inequality Reexamined.* Cambridge, MA: Harvard University Press.

——. 1993. Internal consistency of choice. *Econometrica* 61:495-522.

——. 1999. *Development as Freedom.* New York: Anchor Books.

Shafer, G. 1976. *A Mathematical Theory of Evidence.* Princeton, NJ: Princeton University Press.

Solovay, R. 1970. A model for set theory in which every set of real numbers is Lebesgue measurable. *Annals of Mathematics* 92:1-56.

Stecher, J., and J. Dickhaut. 2008. Generating ambiguity in the laboratory. Available at www.nhh.no/Admin/Public/DWSDownload.aspx?File=%2fFiles%2 fFiler%2finstitutter%2frrr%2fPapers%2fworking+papers%2f01.pdf.

Suppes, P. 1966. Some formal models of grading principles. *Synthese* 6:284-306.

——. 1974. The measurement of belief. *Journal of the Royal Stastical Society* 2: 160-91.

Sutton, J. 2006. Flexibility, profit and survival: in an (objective) model of Knightian uncertainty. Working Paper, Economics Department, London School of Economics.

Temkin, L. 1996. A continuum argument for intransitivity. *Philosophy and Public Affairs* 25:175-210.

Tversky, A. 2003. *Preference, Belief, and Similarity.* Cambridge, MA: MIT Press.

von Mises, R. 1957. *Probability, Statistics, and Truth.* London: Allen and Unwin. (First published in 1928.)

Von Neumann, J., and O. Morgenstern. 1944. *The Theory of Games and Economic Behavior*. Princeton, NJ: Princeton University Press.

Voorhoeve, A. 2007. Heuristics and biases in a purported counterexample to the acyclicity of 'better than'. LSE Choice Group Working Paper.

Wagon, S. 1985. *The Banach–Tarski Paradox*. Cambridge: Cambridge University Press.

Wakker, P. 1989. Continuous subjective expected utility with nonadditive probabilities. *Journal of Mathematical Economics* 18:1–27.

Wald, A. 1938. Die Widersprechsfreiheit des Kollectiv begriffes der Wahrscheinlichkeitsrechnung. *Actualites Scientifiques et Industrielles* 735:79–99.

———. 1950. *Statistical Decision Theory*. New York: Wiley.

Walley, P. 1991. *Statistical Reasoning with Imprecise Probabilities*. London: Chapman & Hall.

Walley, P., and T. Fine. 1982. Towards a frequentist theory of upper and lower probability. *Annals of Statistics* 10:741–61.

Weymark, J. 1991. A reconsideration of the Harsanyi–Sen debate on utilitarianism. In *Interpersonal Comparisons of Well-Being* (ed. J. Elster and J. Roemer). Cambridge: Cambridge University Press.

Index

The Gorman Lectures in Economics
Richard Blundell, Series Editor

Terence (W. M.) Gorman was one of the most distinguished economists of the twentieth century. His ideas are so ingrained in modern economics that we use them daily with almost no acknowledgment. The relationship between individual behavior and aggregate outcomes, two-stage budgeting in individual decision making, the "characteristics" model which lies at the heart of modern consumer economics, and a conceptual framework for "adult equivalence scales" are but a few of these. For over fifty years he guided students and colleagues alike in how best to model economic activities as well as how to test these models once formulated.

During the late 1980s and early 1990s Gorman was a Visiting Professor of Economics at University College London. He became a key part of the newly formed and lively research group at UCL and at the Institute for Fiscal Studies. The aim of this research was to avoid the obsessive labeling that had pigeonholed much of economics and to introduce a free flow of ideas between economic theory, econometrics, and empirical evidence. It worked marvelously and formed the mainstay of economics research in the Economics Department at UCL. These lectures are a tribute to his legacy.

Terence had a lasting impact on all who interacted with him during that period. He was not only an active and innovative economist but he was also a dedicated teacher and mentor to students and junior colleagues. He was generous with his time and more than one discussion with Terence appeared later as a scholarly article inspired by that conversation. He used his skill in mathematics as a framework for his approach but he never insisted on that. What was essential was coherent and logical understanding of economics.

Gorman passed away in January 2003, shortly after the second of these lectures. He will be missed but his written works remain to remind all of us that we are sitting on the shoulders of a giant.

Richard Blundell, University College London and Institute for Fiscal Studies

Biography

Gorman graduated from Trinity College, Dublin, in 1948 in Economics and in 1949 in Mathematics. From 1949 to 1962 he taught in the Commerce Faculty at the University of Birmingham. He held Chairs in Economics at Oxford from 1962 to 1967 and at the London School of Economics from 1967 to 1979, after which he returned to Nuffield College Oxford as an Official Fellow. He remained there until his retirement. He was Visiting Professor of Economics at UCL from 1986 to 1996. Honorary Doctorates have been conferred upon him by the University of Southampton, the University of Birmingham, the National University of Ireland, and University College London. He was a Fellow of the British Academy, an honorary Fellow of Trinity College Dublin and of the London School of Economics, an honorary foreign member of the American Academy of Arts and Sciences and of the American Economic Association. He was a Fellow of the Econometric Society and served as its President in 1972.